# Statistics

Martin Sternstein, Ph.D.

Professor of Mathematics
Ithaca College
Ithaca, New York

BARRON'S

To Faith, Jonathan and Jeremy

*All inquiries should be addressed to:*
Barron's Educational Series, Inc.
250 Wireless Boulevard
Hauppauge, New York 11788

*Library of Congress Catalog Card No. 95-40545*

**International Standard Book No. 0-8120-9311-9**

**Library of Congress Cataloging-in-Publication Data**

Sternstein, Martin.
    Statistics / Martin Sternstein.
        p.    cm. — (*College review series.*  Mathematics)
    Includes index.
    ISBN 0-8120-9311-9
    1. Statistics.    I. Title.    II. Series.
QA276.12.S75   1996
519.5—dc20                                                95-40545
                                                              CIP

PRINTED IN THE UNITED STATES OF AMERICA

987654321

# CONTENTS

## Part 3
# PROBABILITY DISTRIBUTIONS

## Part 4
# THE POPULATION MEAN

# Part 5
# THE POPULATION PROPORTION

# Part 6
# CHI-SQUARE ANALYSIS

# Part 7
# REGRESSION

# APPENDICES

| | |
|---|---|
| Table A | Normal Curve Areas |
| Table B | Critical Values of $t$ |
| Table C | Critical Values of $F$ |
| Table D | The $\chi^2$-Distribution |
| Table E | Critical Levels of $r$ |

# Part 1

# DESCRIPTIVE STATISTICS

*The word statistics comes from the Latin* status, *meaning "state." This derivation reflects that numerical information was historically gathered and interpreted by governments.*

*The ability to describe data in various ways has always been valuable. Now, however, these techniques are becoming more important because of the tremendous, ever-growing number of data being collected in almost all areas of knowledge. The need to make sense out of masses of information has led to the development of formalized ways of describing data. To make intelligent decisions, as well as to understand the world around us, we must become familiar with these primary methods of portraying data.*

*Given a raw set of data, often we can detect no overall pattern. Perhaps some values occur more frequently, a few extreme values may stand out, and usually the range of values is apparent. The presentation of data, including summarizations and descriptions, and involving such concepts as representative or average values, measures of dispersion, positions of various values, and the shape of a distribution, falls under the broad topic of* descriptive statistics, *the subject of Part 1 of this book. This aspect of statistics is in contrast to* analysis, *the process of drawing inferences from limited data, a subject that is discussed in later chapters.*

*More specifically, Chapters 1–4 are concerned with (1) how to measure central tendency; (2) how to measure variability; (3) how to measure position; and (4) how to display shape.*

# 1
# CENTRAL TENDENCY

The word *average* is used in phrases common to everyday conversation. People speak of bowling and batting averages or the average life expectancy of a battery or a human being. Actually the word *average* is derived from the French *avarie*, which refers to the money shippers used to contribute to help compensate for the losses suffered by other shippers whose cargo did not arrive safely (that is, the losses were shared, with everyone contributing an *average* amount). In common usage *average* has come to mean a *representative* score or a *typical* value or the *center* of a distribution. Mathematically, there are a variety of ways to define the average of a set of data. In practice, we use whichever method is most appropriate for the particular case under consideration.

In the following paragraphs we consider the three primary ways of denoting an average:

1. The *median*, which is the middle number of a set of numbers arranged in numerical order.
2. The *mode*, which is the number that occurs most frequently.
3. The *mean*, which is found by summing items in a set and then dividing by the number of items.

**Example 1.1** Consider the following set of homerun distances (in feet) to center field in 13 ballparks: {387, 400, 400, 410, 410, 410, 414, 415, 420, 420, 421, 457, 461}. What is the average?

**Answer:** The median is 414 (there are six values below 414 and six values above). The mode is 410 (occurs three times). The mean is

$$\frac{387 + 400 + 400 + 410 + \ldots + 457 + 461}{13} = 417.3 \; feet$$

# MEDIAN

The word *median* is derived from the Latin *medius* which means "middle." The values under consideration are arranged in ascending or descending order. If there is an odd number of values, the median is the middle one. If there are an even number, the median is found by adding the two middle values and dividing by 2. Thus the median of a set has the same number of elements above it as there are below it.

The median is not affected by exactly how large the larger values are or by exactly how small the smaller values are. Thus it is a particularly useful measurement when the extreme values, called *outliers*, are in some way suspicious, or when we want to diminish their effect. For example, if ten mice try to solve a maze, and nine succeed in less than 15 minutes, while one is still trying after 24 hours, the most representative value is the median (not the mean, which is over 1 hour). Similarly, if the salaries of four executives are each between $240,000 and $245,000 while a fifth is paid less than $20,000, again the most representative value is the median (the mean is under $200,000).

In certain situations the median offers the most economical and quickest way to calculate an average. For example, suppose 10,000 light bulbs of a particular brand are installed in a factory. An average life expectancy for the bulbs can most easily be found by noting how much time passes before exactly one-half of them have had to be replaced. The median is also useful in certain kinds of medical research. For example, to compare the relative strengths of different poisons, a scientist notes what dosage of each poison will result in the deaths of exactly one-half the test animals. If one of the animals proves especially susceptible to a particular poison, this median lethal dose is not affected.

# MODE

The *mode*, or most frequent value, is an easily understood representative score. It is clear what is meant by "The most common family size is four" or "That professor gives out more Bs than any other grade." If exactly one value must be chosen as a basis for certain decisions, this value is often the mode. For example, if a supplier can expand production of only a single item, he may well decide upon the one most frequently sold. However, since the mode presents the typical result, its use tends to rule out further arithmetic calculations based on this result. For example, knowing that a certain player is the most popular with the fans does not lead to calculations based simply on this knowledge.

When two numbers in a set have equal frequency, and this frequency is higher than any other, we say that the set is *bimodal.* Even when one frequency is greater than the other, but both are above the other frequencies, the set will sometimes be termed bimodal. Occasionally, the values in a set are distributed so evenly that the idea of a modal value has very little meaning for that set.

# MEAN

While the median and mode are often useful in *descriptive* statistics, the *mean* or, more accurately, the *arithmetic mean*, is most important for statistical *inference and analysis*. Also, for the layperson, the "average" is usually understood to be the "mean."

The mean of a *whole population* (the complete set of items of interest) is often denoted by the Greek letter $\mu$ (mu), while the mean of a *sample* (a part of a population) is often denoted by $\bar{x}$. For example, the mean value of the set of all houses in the United States might be $\mu$ = $56,400, while the mean value of 100 randomly chosen houses might be $\bar{x}$ = $52,100 or perhaps $\bar{x}$ = $63,800 or even $\bar{x}$ = $124,000.

In statistics we learn how to estimate a population mean from a sample mean. Throughout this book, the word *sample* implies a *random* sample, that is, a sample selected in such a way that each element of the population has an equal chance to be included in the sample. In the real world, this process of *random* selection is often very difficult to achieve.

Mathematically, the mean = $(\Sigma x)/n$, where $\Sigma x$ represents the sum of all the elements of the set under consideration, and $n$ is the actual number of elements. $\Sigma$ is the uppercase Greek letter sigma.

Unlike the mode or median, the mean is sensitive to a change in any value, as shown in Example 1.2.

**Example 1.2**   Suppose that the numbers of unnecessary procedures recommended by five doctors in a 1-month period are given by the set {2, 2, 8, 20, 33}. Note that the median is 8, the mode is 2, and the mean is $\dfrac{2 + 2 + 8 + 20 + 33}{5}$ = 13. If it is discovered that the fifth doctor recommended an additional 25 unnecessary procedures, how are the median, mode and mean affected?

*Answer:* The set is now {2, 2, 8, 20, 58}. The median is still 8 and the mode is still 2. The mean, however, changes to $\dfrac{2 + 2 + 8 + 20 + 58}{5}$ = 18.

Adding the same constant to each value will increase the mean by a like amount. Similarly, multiplying each value by the same constant will multiply the mean by a like amount. (See Study Questions 1 and 2.)

**Example 1.3**   Suppose the salaries of six employees are $3000, $7000, $15,000, $22,000, $23,000, and $38,000.
*a.* What is the mean salary?

*Answer:*

$$\frac{3000 + 7000 + 15,000 + 22,000 + 23,000 + 38,000}{6} = \$18,000.$$

b. What will the new mean salary be if everyone receives a $3000 increase?

**Answer:**

$$\frac{6000 + 10{,}000 + 18{,}000 + 25{,}000 + 26{,}000 + 41{,}000}{6} = \$21{,}000.$$

Note that $18{,}000 + \$3000 = \$21{,}000$.

c. What if everyone receives a 10% raise?

**Answer:**

$$\frac{3300 + 7700 + 16{,}500 + 24{,}200 + 25{,}300 + 41{,}800}{6} = \$19{,}800.$$

Note that 110% of $18{,}000$ is $\$19{,}800$.

Because many real-life applications of statistics involve comparisons of *two* populations, it is important for later use to note how the mean is calculated for a set of differences.

**Example 1.4** Suppose set X = {2, 9, 11, 22} and set Y = {5, 7, 15}. Note that the mean of set X is $\mu_x = \frac{2 + 9 + 11 + 22}{4} = 11$ and the mean of set Y is $\mu_y = \frac{5 + 7 + 15}{3} = 9$. Form the set Z of differences by subtracting each element of Y from each of X:

Z = {2−5, 2−7, 2−15, 9−5, 9−7, 9−15, 11−5, 11−7, 11−15, 22−5, 22−7, 22−15}
= {−3, −5, −13, 4, 2, −6, 6, 4, −4, 17, 15, 7}

What is the mean of Z?

**Answer:**

$$\mu = \frac{-3 - 5 - 13 + 4 + 2 - 6 + 6 + 4 - 4 + 17 + 15 + 7}{12}$$
$$= \frac{24}{12} = 2$$

Note that $\mu_z = \mu_x - \mu_y$

In general, the mean of a set of differences is equal to the difference of the means of the two original sets. Even more generally, if a sum is formed by picking one element from each of several sets, the mean of such sums is simply the sum of the means of the various sets.

Three secondary procedures for denoting an average are the *harmonic mean*, the *geometric mean*, and the *trimmed mean*.

# HARMONIC MEAN

The *harmonic mean* of a set of $n$ numbers is defined to be

$$H = \frac{n}{\sum \frac{1}{x}}.$$

that is, $n$ divided by the sum of the reciprocals of the numbers.

**Example 1.5** Suppose four kinds of nails cost $0.01, $0.02, $0.03, and $0.04 apiece. What is the average cost of a nail if we purchase $3.00 worth of each kind?

*Answer:* The number of $0.01 nails purchased is 3.00/0.01 = 300. Similarly, the numbers of $0.02, $0.03, and $0.04 nails purchased are 3.00/0.02 = 150, 3.00/0.03 = 100, and 3.00/0.04 = 75, respectively. The total number of nails purchased is 300 + 150 + 100 + 75 = 625, while the total amount spent is 4($3.00) = $12.00. Thus the average cost per nail is 12.00/625 = $.0192.

Note that the answer could have been calculated immediately as the *harmonic mean* of the set {0.01, 0.02, 0.03, 0.04}:

$$\frac{4}{\frac{1}{0.01} + \frac{1}{00.02} + \frac{1}{0.03} + \frac{1}{0.04}} = 0.0192$$

Also note that if we bought 100 of each of the four kinds of nails (that is, same number of each kind rather than same money spent on each), then the average spent per nail would be (.01 + .02 + .03 + .04)/4 = $.025.

# GEOMETRIC MEAN

The *geometric mean* of a set {$y_1, y_2,..., y_n$} is defined to be

$$\sqrt[n]{y_1 y_2 \cdots y_n},$$

that is, the $n$th root of the product of the elements. In particular, the geometric mean of two numbers is the square root of their product. The geometric mean is appropriate in situations of what is known as geometric growth.

**Example 1.6**    The population of the United States in 1820 was 9,638,453 and in 1840 was 17,069,453. Estimate the population in 1830 using the geometric mean.

*Answer:*

$$\sqrt{(9,638,453)(17,069,453)} = 12,826,657$$

(The actual value was 12,866,020.)

# TRIMMED MEAN

The *trimmed mean* represents an attempt to reduce the influence of extreme values (outliers) on the mean. It is calculated by arranging the terms in numerical order, deleting the first quarter and the fourth quarter of the values, and then finding the arithmetic mean of what remains. (Alternatively, instead of 25%, any percent, for example 5% or 10%, can be taken away from both the upper and lower ends, depending on the makeup of the particular set under consideration.)

**Example 1.7**    The World Bank has predicted that the 20 most populous countries in the year 2100 and their populations (in millions) will be as follows:

India, 1632; China, 1571; Nigeria, 509; Russia, 376; Indonesia, 356; Pakistan, 316; United States, 309; Bangladesh, 297; Brazil, 293; Mexico, 196; Ethiopia, 173; Vietnam, 168; Iran, 164; Zaire, 139; Japan, 128; Philippines, 125; Tanzania, 120; Kenya, 116; Burma, 112; and Egypt, 111. Calculate the mean and the trimmed mean for the average predicted populations of these countries.

*Answer:* The mean is $(1632 + ... + 111)/20 = 361$ million, while the trimmed mean, without the extreme influence of India, China, and Nigeria is $(316 + ... + 128)/10 = 218$ million.

# EXERCISES

1. In the second game of the 1989 World Series between Oakland and San Francisco (played 2 days before the northern California earthquake postponed the series), ten players went hitless, eight players had one hit apiece, and one player (Rickey Henderson) had three hits. What was the mean, median, and mode number of hits?

2. In 1993 the eight countries in which the most patents were filed were as follows: Japan, 208,347; United States, 57,890; Germany, 42,922; France,

11,187; Britain, 9333; China, 5566; Italy, 3726; and Australia, 3315 (*The Economist*, August 27, 1994). What are the median, mean, and trimmed mean averages for the number of 1993 patents in these eight countries?

3. On August 23, 1994, the 5-day total returns for equal investments in foreign stocks, U.S. stocks, gold, money market funds, and treasury bonds averaged 0.086% (*Business Week*, September 5, 1994). If the respective returns for foreign stocks, U.S. stocks, gold, and money market funds were +0.70%, −0.11%, +1.34%, and +0.05%, what was the return for treasury bonds?

4. The weights (in pounds) of the members of the 1968 Chicago Bear offensive line were as follows: Bob Pickens, 255; Jim Cadile, 240; Mike Pyle, 250; George Seals, 270; Rufus Mayes, 260. What was the average (mean) weight of these linesmen? What would the average weight have been if each of the players had gained 10 pounds? If each had increased his weight by 10 percent?

5. Fecal pollution can lead to illness among swimmers in small lakes. The Environmental Protection Agency recommends that enterococcus counts at freshwater swimming areas not exceed a geometric mean of 33 per deciliter (EPA report no. 440/5-84-002). If three samples of lake water taken at different sections of a swimming area show pathogen counts of 25, 30, and 40 organisms per deciliter, determine whether the geometric mean exceeds the EPA limit.

6. The results of a recent University of Chicago study reported in *The New York Times* are tabulated below:

### SEX PARTNERS SINCE AGE 18

|       | 0  | 1   | 2–4 | 5–10 | 11–20 | 21+ |
|-------|----|-----|-----|------|-------|-----|
| Men   | 3% | 20% | 21% | 23%  | 16%   | 17% |
| Women | 3% | 32% | 36% | 20%  | 6%    | 3%  |

In which category does the median number of partners fall for men? For women? Explain.

7. In your daily newspaper, find examples of two sets of numbers, in one of which the mean is greater than the median, and in the other the mean is less than the median. Can you find an example of data where the mean and median are equal?

## Study Questions

1. Show that the addition of the same constant to each value in a set will increase the mean by a like amount. [Hint: suppose that the

set has the $n$ elements: $x_1$, $x_2$, $x_3$, ..., $x_n$. Then the mean is: $\mu = \dfrac{x_1 + x_2 + x_3 + \ldots + x_n}{n}$. Show that the mean of the set $\{x_1 + c, x_2 + c, x_3 + c, ..., x_n + c\}$ is $\mu + c$.]

2. Prove that multiplying each value in a set by the same constant will multiply the mean by a like amount.

3. Questions 1 and 2 form the background for a technique called *coding* which in certain cases provides a shortcut for computing the mean. Using these addition and multiplication principles, how would you mentally calculate the mean of the set $\{25.002, 25.015, 25.013, 25.008\ 25.009, 25.001\}$?

4. Suppose $X = \{x_1, x_2, \ldots, x_m\}$ and $Y = \{y_1, y_2, \ldots, y_n\}$. Form the set of differences $Z = \{x_i - y_j\}$, $i = 1, 2, ..., m$ and $j = 1, 2, ..., n$. Show that $\mu_z = \mu_x - \mu_y$.

$$[\textit{Hint:}\ \Sigma z\ =\ n\,\Sigma x - m\,\Sigma y. \qquad \textit{Why?} \qquad \textit{Then}\ \frac{\Sigma z}{mn} = ?]$$

# 2
# VARIABILITY

In describing a set of numbers, not only is it useful to designate an average value, but also it is important to be able to indicate the *variability* or the *dispersion* of the measurements. A producer of time bombs aims for small variability—it would not be good if his 30 minute fuses actually had a range of 10 to 50 minutes before detonation. On the other hand, a teacher interested in distinguishing the better from the poorer students aims to design exams with large variability in results—it would not be helpful if all her students scored exactly the same. The players on two basketball teams may have the same average height, but this fact doesn't tell the whole story. If the dispersions are quite different, one team may have a 7-foot player, whereas the other has no one over 6 feet tall. Two Mediterranean holiday cruises may advertise the same average age for their passengers. One, however, may have only passengers between 20 and 25 years old, while the other has only middle-aged parents in their 40s together with their children under age 10.

There are four primary ways of describing variability or dispersion:

1. The *range*, which is the difference between the largest and smallest values.
2. The *average deviation*, which is found by averaging the absolute differences of all the values from the mean.
3. The *variance*, which is determined by averaging the squared differences of all the values from the mean.
4. The *standard deviation*, which is the square root of the variance.

**Example 2.1** The monthly rainfall in Monrovia, Liberia, where May through October is the so-called rainy season and November through April the dry season, is as follows:

| Month: | Jan | Feb | Mar | Apr | May | Jun | Jul | Aug | Sep | Oct | Nov | Dec |
|---|---|---|---|---|---|---|---|---|---|---|---|---|
| Rain (in.): | 1 | 2 | 4 | 6 | 18 | 37 | 31 | 16 | 28 | 24 | 9 | 4 |

The mean is

$$\frac{1 + 2 + 4 + 6 + 18 + 37 + 31 + 16 + 28 + 24 + 9 + 4}{12}$$
$$= 15 \ inches$$

What are the measures of variability?

**Answer:** Range: The maximum is 37 inches (June), and the minimum is 1 inch (January). Thus the range is $37 - 1 = 36$ inches of rain.

Average deviation:

$$\frac{|1 - 15| + |2 - 15| + |4 - 15| + |6 - 15| + |18 - 15| + \dots + |4 - 15|}{12}$$

$$= \frac{14 + 13 + 11 + 9 + 3 + 22 + 16 + 1 + 13 + 9 + 6 + 11}{12} = 10.7 \text{ inches}$$

Variance:

$$\frac{14^2 + 13^2 + 11^2 + 9^2 + 3^2 + 22^2 + 16^2 + 1^2 + 13^2 + 9^2 + 6^2 + 11^2}{12} = 143.7$$

Standard deviation: $\sqrt{143.7} = 12.0 \text{ inches}$

# RANGE

The simplest, most easily calculated, measure of variability is the *range*. The difference between the largest and smallest values can be noted quickly, and it does give some impression of the dispersion. However, it is entirely dependent on the two extreme values and is insensitive to the ones in the middle.

One use of the range is to evaluate samples with very few items. For example, some quality control techniques involve taking periodic small samples and basing further action upon the range found in several such samples.

# AVERAGE DEVIATION

Unlike the range, most other measures of variability are concerned with the dispersion around some center or representative value. The *average deviation* is the sum of the deviations from the mean (without regard to sign) divided by the number of elements:

$$Average\ deviation = \frac{\Sigma |x - \mu|}{n}$$

where $\mu$ is the mean and $n$ is the population size.

The average deviation gives information in an easily understood form. If a town has a mean yearly snowfall of 90 inches with an average deviation of 10 inches, the reader has an intuitive sense that a yearly snowfall between 80 and 100 inches would probably not be unusual. If two companies advertise mean starting salaries of $25,000, but in one the average deviation is $500 while in the other it is $8,000, applicants will have some feeling for what this difference means. The average deviation is also useful with regard to discrepancies in populations where all the values are planned rather than related to chance occurrence. In such cases it is worthwhile to discuss how far from the mean each value is chosen to be, and what the average is of these differences.

# VARIANCE

In many situations, dispersion is not planned, but rather is the result of various chance happenings. In such cases, the average deviation is not the proper tool for measuring variability. For example, consider the motion of microscopic particles suspended in a liquid. The unpredictable motion of any particle is the result of many small movements in various directions caused by random bumps from other particles. If we average the total displacements of all the particles from their starting points (this corresponds to the average deviation), the result does not increase in direct proportion to time. If, however, we average the *squares* of the total displacements of all the particles, this result does increase in direct proportion to time.

The same holds true for the movement of paramecia. Their seemingly random motions as noted under a microscope can be described by the fact that the average of the *squares* of the displacements from their starting points is directly proportional to time. Also, consider ping-pong balls dropped straight down from a high tower and subjected to chance buffeting in the air. We can measure the deviations from a center spot on the ground to the spots where the balls actually strike. As the height of the tower is increased, the average of the *squared* deviations increases proportionately.

In a wide variety of cases we are in effect trying to measure dispersion from the mean due to a multitude of chance effects. The proper tool in these cases is the average of the squared deviations from the mean. This is called the *variance* and is denoted by $\sigma^2$ ($\sigma$ is the lower case Greek letter sigma):

$$\sigma^2 = \frac{\Sigma(x - \mu)^2}{n}$$

For circumstances specified in Chapter 11, the variance of a sample, denoted by $s^2$, is calculated as

$$s^2 = \frac{\Sigma(x - \overline{x})^2}{n - 1}$$

**Example 2.2**    During the years 1929 through 1939 of the Great Depression, the weekly average hours worked in manufacturing jobs were 45, 43, 41, 39, 39, 35, 37, 40, 39, 36, and 37. What is the variance?

**Answer:**

$$\mu = \frac{45 + 43 + 41 + 39 + 39 + 35 + 37 + 40 + 39 + 36 + 37}{11}$$

$$= 39.2 \; hours$$

$$\sigma^2 = \frac{(45 - 39.2)^2 + (43 - 39.2)^2 + \ldots + (36 - 39.2)^2 + (37 - 39.2)^2}{11} = 8.1$$

Important for later use is a procedure to calculate the variance for a set of differences.

**Example 2.3**    Consider the following from Chapter 1:

$X = \{2, 9, 11, 22\}$, $Y = \{5, 7, 15\}$, and $Z = $ the set of differences $\{-3, -5, -13, 4, 2, -6, 6, 4, -4, 17, 15, 7\}$. What is the variance of $Z$?

**Answer:**

$$\mu_x = 11, \sigma_x^2 = \frac{(2 - 11)^2 + (9 - 11)^2 + (11 - 11)^2 + (22 - 11)^2}{4}$$

$$= \frac{206}{4} = 51.5$$

$$\mu_y = 9, \sigma_y^2 = \frac{(5 - 9)^2 + (7 - 9)^2 + (15 - 9)^2}{3} = \frac{56}{3} = 18.67$$

$$\mu_z = 2,$$

$$\sigma_z^2 = \frac{(-3 - 2)^2 + (-5 - 2)^2 + (-13 - 2)^2 + \ldots + ((7 - 2)^2}{12}$$

$$= \frac{842}{12} = 70.17$$

How are $\sigma_x^2, \sigma_y^2$ *and* $\sigma_z^2$ related? Note that in the above example, $\sigma_z^2 = \sigma_x^2 + \sigma_y^2$! This is true for the variance of any set of differences. More generally, if a total is formed by a procedure that adds or subtracts one element from each of several sets, the variance of the resulting totals is simply the sum of the variances of the several sets.

Not only can we sum variances as shown above to calculate total variance, but we can also reverse the process and determine how the total variance is split up among its various sources. For example, we can find what portion of the variance in numbers of sales by a company's sales representatives is due to the individual salesperson, what portion is due to the territory, what portion is due to the particular products sold by each salesperson, and so on. (The average deviation cannot be partitioned in this way.)

An arithmetic tool for calculating the variance is:

$$\sigma^2 = \frac{\sum x^2}{n} - \mu^2$$

In words, the variance can be found by subtracting the square of the mean from the average of the squared scores.

**Example 2.4**    Let $X = \{3, 7, 15, 23\}$. What is the variance?

*Answer:*

$$\Sigma x = 3 + 7 + 15 + 23 = 48$$

$$\Sigma x^2 = 3^2 + 7^2 + 15^2 + 23^2 = 812$$

$$\mu = \frac{\Sigma x}{n} = \frac{48}{4} = 12$$

The variance can be calculated from its definition:

$$\sigma^2 = \frac{\Sigma(x - \mu)^2}{n}$$

$$= \frac{(3 - 12)^2 + (7 - 12)^2 + (15 - 12)^2 + (23 - 12)^2}{4} = 59$$

Or it can be calculated as follows:

$$\sigma^2 = \frac{\Sigma x^2}{n} - \mu^2 = \frac{812}{4} - 12^2 = 203 - 144 = 59$$

Similarly, there is an arithmetical tool for calculating the variance of a sample:

$$s^2 = \frac{\Sigma x^2 - \frac{(Sx)^2}{n}}{n - 1}$$

# STANDARD DEVIATION

Suppose we wish to pick a representative value for the variability of some population. The preceding discussions indicate that a natural choice is the value whose *square* is the average of the *squared* deviations from the mean. Thus we are led to consider the *square root* of the variance. This is called the *standard deviation*, is denoted by $\sigma$, and is calculated as follows:

$$\sigma = \sqrt{\frac{\Sigma(x - \mu)^2}{n}} = \sqrt{\frac{\Sigma x^2}{n} - \mu^2}$$

Similarly, the standard deviation of a sample (see Chapter 11) is denoted by *s* and is calculated as follows:

$$\sigma = \sqrt{\frac{\Sigma(x - \overline{x})^2}{n - 1}} = \sqrt{\frac{\Sigma x^2 - \dfrac{(\Sigma x)^2}{n}}{n - 1}}$$

**Example 2.5**  The approximate numbers of bank suspensions during the years 1928 through 1933 were 500, 650, 1350, 2300, 1450, and 4000, respectively. What is the standard deviation?

***Answer:***

$$\mu = \frac{500 + 650 + 1350 + 2300 + 1450 + 4000}{6} = 1700$$

$$\sigma^2 = \frac{1200^2 + 1050^2 + 375^2 + 575^2 + 250^2 + 2300^2}{6} = 1,394,375$$

$$\sigma = \sqrt{1,394,375} \approx 1181$$

Two secondary procedures for measuring variability or dispersion are the *interquartile range* and the *relative variability.*

# INTERQUARTILE RANGE

This is one method of removing the influence of extreme values on the range. The *interquartile range* is calculated by arranging the data in numerical order, removing the upper and lower one-quarter of the values, and then noting the range of the remaining values.

**Example 2.6**  Suppose that farm sizes in a small community have the following characteristics: the smallest value is 16.6 acres, 10% of the values are below 23.5 acres, 25% are below 41.1 acres, the median is 57.6 acres, 60% are below 87.2 acres, 75% are below 101.9 acres, 90% are below 124.0 acres, and the top value is 201.7 acres.
*a.* What is the range?

***Answer:*** The range is 201.7 − 16.6 = 185.1 acres.

*b.* What is the interquartile range?

***Answer:*** The interquartile range, with the highest and lowest one-quarter of the values removed, is 101.9 − 41.1 = 60.8 acres. Thus, while the largest farm is 185 acres more than the smallest, the middle 50% of the farm sizes range over a 61 acre interval.

# RELATIVE VARIABILITY

A comparison of two variances is more meaningful if the means of the populations are also taken into consideration. *Relative variability* is defined to be the quotient obtained by dividing the standard deviation by the mean. Usually the result is then expressed as a percentage.

**Example 2.7**    Suppose that the mean salary for police officers is $45,000 with a standard deviation of $9000, while the mean salary for fire fighters is $35,000 with a standard deviation of $7700. What are the respective relative variabilities?

*Answer:* The relative variabilities are 9000/45,000 and 7700/35,000, or 20% and 22%, respectively. Thus, while the police officers' salaries vary more in absolute terms, the fire fighters' salaries vary more in relative terms when the difference in means is taken into consideration.

# EXERCISES

1. The number of criminals executed in the United States during the 1930s and 1940s were as follows: 1930—155; 1931—153; 1932—140; 1933—160; 1934—168; 1935—199; 1936—195; 1937—147; 1938—190; 1939—160; 1940—124; 1941—123; 1942—147; 1943—131; 1944—120; 1945—117; 1946—131; 1947—153; 1948—119; 1949—119. What are the mean, the range, and the standard deviation?

2. According to a 1988 *New York Times* article, the ten car models with the highest theft rates were as follows:

| Vehicle Model | Thefts per 1000 Cars |
|---|---|
| Pontiac Firebird | 30.14 |
| Chevrolet Camaro | 26.02 |
| Chevrolet Monte Carlo | 20.28 |
| Toyota MR2 | 19.25 |
| Buick Regal | 14.70 |
| Mitsubishi Starion | 14.70 |
| Ferrari Mondial | 13.60 |
| Mitsubishi Mirage | 12.80 |
| Pontiac Fiero | 12.68 |
| Oldsmobile Cutlass | 11.73 |

a. What are the mean, range, and standard deviation of these theft rates?
b. How will each of these values change if each theft rate increases by 1.5? by 15%?

3. Because Brazilian doctors do not use generic prescriptions, black markets exist in Brazil for many drugs (*The Lancet*, October 15, 1994). For example, the black market price per tablet of acetaminophen for seven brands are as follows: Acetofen, $0.23; Dorico, $0.10; Pacemol, $0.17; Parador, $0.14; Tylenol, $0.18; Paracetamol, $0.08; and Nevralgina, $0.27. What are the mean, range, and standard deviation of these prices?

## Study Questions

1. Suppose that the same constant is added to every value in a population. How does this affect the range? The average deviation? The variance? The standard deviation?

2. Suppose that each value in a set is multiplied by the same constant. How does this affect the range? The average deviation? The variance? The standard deviation?

3. Does it make sense to talk about the average of the deviations from the mean (*with* regard to sign)? In other words, what can be said about
$$\frac{\Sigma(x - \mu)}{n} ?$$

4. If the median was the most significant measure of central tendency, the appropriate measure of variability might well be $\dfrac{\Sigma |x - M|}{n}$. Show that $\Sigma |x - K|$ is smallest when $K = M$ (the median). [*Hint*: Think about the position of $M$ among the $x$s. As we move away from $M$ by a small amount $k$, what happens to each $|x - M|$ in changing to $|x - K|$ where $K = M + k$]

5. If you've had an introductory calculus course, use calculus to show that $\Sigma(x - t)^2$ is smallest when $t = \mu$. [*Hint*: $\dfrac{d}{dt} \Sigma(x - t)^2 = ?$ Set equal to 0 and solve for $t$.]

# 3
# POSITION

We have seen several ways of choosing a value to represent the center of a distribution. We also need to be able to talk about the *position* of any other value. In some situations, such as wine tasting, simple rankings are of interest. Other cases, for example, evaluating college applications, may involve positioning according to percentile rankings. There are also situations in which position can be specified by making use of measurements of both central tendency and variability.

There are three important, recognized procedures for designating position:

1. *Simple ranking*, which involves arranging the elements in some order and noting where in that order a particular value falls.
2. *Percentile ranking*, which indicates what percent of all values fall below the value under consideration.
3. The *z-score*, which tells very specifically by how many standard deviations a particular value varies from the mean.

**Example 3.1** The water capacities (in gallons) of the 57 major solid-fuel boilers sold in the United States are as follows: 6.3, 7.4, 8.6, 10, 12.1, 50, 8.2, 9.8, 11.4, 12.9, 14.5, 16.1, 26, 21, 27, 40, 55, 30, 35, 55, 65, 18.8, 23.3, 28.3, 33.8, 26.4, 33, 50, 35, 21, 21, 18.5, 26.4, 37, 12, 12, 50, 65, 65, 56, 66, 60, 70, 27.7, 34.3, 42, 46.2, 33, 25, 29, 40, 19, 9, 12.5, 15.8, 24.5, and 16.5 (John W. Bartok, *Solid-Fuel Furnaces & Boilers*, Garden Way). What is the position of the Passat HO-45 which has a capacity of 46.2 gallons?

*Answer:* Since there are 12 boilers with higher capacities on the list, the Passat has a *simple ranking* of 13th (out of 57).

Forty-four boilers have lower capacities, so the Passat has a *percentile ranking* of 44/57 = 77.2%.

The above list has a mean of 30.2 and a standard deviation of 17.8, so the Passat has a *z-score* of (46.2 − 30.3)/17.8 = 0.89.

# SIMPLE RANKING

*Simple ranking* is easily calculated and easily understood. We know what it means for someone to graduate second in a class of 435, or for a player from a team of size 30 to have the seventh best batting average. Simple ranking is useful even when no numerical values are associated with the elements. For example, detergents may be ranked according to relative cleansing ability without any numerical measurements of strength.

# PERCENTILE RANKING

*Percentile ranking*, another readily understood measurement of position, is helpful for comparing positions with different bases. We can more easily compare a rank of 176 out of 704 with a rank of 187 out of 935 by noting that the first has a rank of 75%, the second a rank of 80%. Percentile rank is also useful when the exact population size is not known or is irrelevant. For example, it is more meaningful to say that Jennifer scored in the 90th percentile on a national exam, rather than trying to determine her exact ranking among some large number of test takers.

# Z-SCORE

The *z-score* is a measure of position that takes into account both the center and the dispersion of the distribution. More specifically, the *z*-score of a value tells how many standard deviations the value is from the mean. Mathematically, $x - \mu$ gives the raw distance from $\mu$ to $x$; dividing by $\sigma$ converts this to numbers of standard deviations. Thus $z = (x - \mu)/\sigma$, where $x$ is the raw score, $\mu$ is the mean, and $\sigma$ is the standard deviation. If the score $x$ is greater than the mean $\mu$, then z is positive; if $x$ is less, then $z$ is negative.

Given a *z*-score, we can reverse the procedure and find the corresponding raw score. Solving for $x$ gives $x = \mu + z\sigma$.

**Example 3.2**    Suppose that the average (mean) price of gasoline in a large city is $1.80 per gallon with a standard deviation of $0.05. Then $1.90 has a *z*-score of $(1.90 - 1.80)/0.05 = +2$, while $1.65 has a *z*-score of $(1.65 - 1.80)/0.05 = -3$. Alternatively, a *z*-score of $+2.2$ corresponds to a raw score of $1.80 + 2.2(0.05) = 1.80 + 0.11 = 1.91$, while a *z*-score of $-1.6$ corresponds to $1.80 - 1.6(0.05) = 1.72$.

It is often useful to portray integer *z*-scores and the corresponding raw scores as follows:

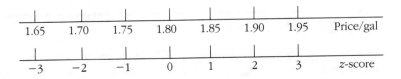

**Example 3.3**    Suppose the attendance at a movie theater averages 780 with a standard deviation of 40. Adding and subtracting multiples of 40 to the mean 780 gives:

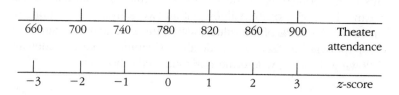

An attendance of 835 would convert to a z-score as follows:

$$\frac{835 - 780}{40} = \frac{55}{40} = 1.375$$ A z-score of $-2.15$ would convert to a theater attendance as follows: $780 - 2.15(40) = 694$.

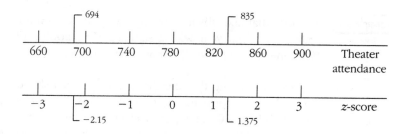

**Example 3.4**    An assembly line produces an average of 12,600 units per month with a standard deviation of 830 units. Adding and subtracting multiples of 830 to the mean, 12,600, gives:

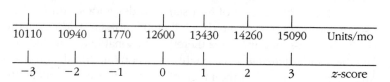

Thus 11,106 has a z-score of $(11,106 - 12,600)/830 = -1.8$, while a z-score of 2.4 corresponds to $12,600 + 2.4(830) = 14,592$.

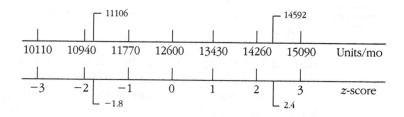

# EMPIRICAL RULE

The *empirical rule* applies specifically to symmetric, "bell-shaped" data. In this case, about 68% of the values lie within one standard deviation of the mean; about 95% of the values, within two standard deviations of the mean; and more than 99% of the values, within three standard deviations of the mean.

Shape will be discussed in detail in Chapter 4, but consider intuitively the following displays, where the horizontal axis shows $z$-scores:

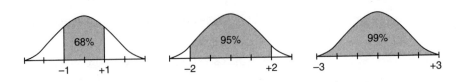

**Example 3.5**  Suppose that taxicabs in New York City are driven an average of 75,000 miles per year with a standard deviation of 12,000 miles. What information does the empirical rule give us?

*Answer:* Assuming that the distribution is "bell-shaped," we can conclude that approximately 68% of the taxis are driven between 63,000 and 87,000 miles per year, approximately 95% are driven between 51,000 and 99,000 miles, and virtually all are driven between 39,000 and 111,000 miles.

The empirical rule also gives a useful quick estimate of the standard deviation in terms of the range. We can see in the display above that 95% of the data fall within a span of four standard deviations (from −2 to +2 on the $z$-score line), and 99% of the data fall within six standard deviations (from −3 to +3 on the $z$-score line). It is therefore reasonable to conclude that for these data the standard deviation is roughly between one-fourth and one-sixth of the range. Since we can find the range of a set almost immediately, the empirical rule technique for estimating the standard deviation is often helpful in pointing out probable arithmetic errors.

**Example 3.6**  If the range of a data set is 60, what is an estimate for the standard deviation?

*Answer:* By the empirical rule, the standard deviation would be expected to be between $(1/6)60 = 10$ and $(1/4)60 = 15$. If the standard deviation is calculated to be 0.32 or 87 there is a probable arithmetic error; a calculation of 12, however, is reasonable.

# CHEBYSHEV'S THEOREM

When data are spread out, the standard deviation is larger; when data are tightly compacted, the standard deviation is smaller. However, no matter what the dispersion, and even if the data are not bell-shaped, certain percentages of the data will always fall within specified numbers of standard deviations from the mean.

The Russian mathematician Chebyshev showed that, for any set of data, *at least* $1 - 1/k^2$ of the values lie within $k$ standard deviations of the mean. This result is known as *Chebyshev's Theorem.* In terms of percentages, at least $100(1 - 1/k^2)\%$ of the values are within $k$ standard deviations of the mean. In terms of $z$-scores, at least $1 - 1/k^2$ of the values have $z$-scores between $-k$ and $+k$. Therefore, when $k = 3$, at least $1 - 1/9$ or 88.89% of the values lie within three standard deviations of the mean. And, when $k = 5$, at least $1 - 1/25$ or 96% of the values have $z$-scores between $-5$ and $+5$.

**Example 3.7**  Suppose an electronic part takes an average of 3.4 hours to move through an assembly line with a standard deviation of 0.5 hour. What does Chebyshev's Theorem give using $k = 2$ and $k = 4$?

*Answer:* Using $k = 2$, we find that at least $1 - 1/4$ or 75% of the parts take between $3.4 - 2(0.5) = 2.4$ hours and $3.4 + 2(0.5) = 4.4$ hours to move through the line. Similarly, with $k = 4$, at least 15/16 or 93.75% of the parts take between 1.4 and 5.4 hours.

Given a range around the mean, we can convert to $z$-scores and ask about the percentage of values in the range. If one set has mean $\mu = 85$ and standard deviation $\sigma = 1$, while a second set has $\mu = 85$ and $\sigma = 5$, then, for example, the percentages of values between 75 and 95 will be different. For the first set, the relevant $z$-scores are $\pm 10$; for the second the relevant $z$-scores are $\pm 2$. Thus at least 99/100 or 99% of the first set's values, but possibly only 3/4 or 75% of the second set's values, lie between 75 and 95.

**Example 3.8**  Suppose the daily intake at a toll booth averages \$3500 with a standard deviation of \$200. What percentage of the daily intakes should be between \$3000 and \$4000?

*Answer:* The relevant $z$-scores are $\pm 500/200 = \pm 2.5$, so Chebyshev's theorem says that at least $1 - 1/(2.5)^2 = 21/25$ or 84% of the daily intakes should be between \$3000 and \$4000.

Note that, for $k = 1$, $1 - 1/k^2 = 0$; here the theorem gives no useful information.

**Reminder:** The power of Chebyshev's theorem is that it applies to *all* sets of data. However, if the data are bell-shaped, we can draw stronger conclusions by using the empirical rule stated earlier.

# EXERCISES

1. According to the National Center for Health Statistics, of the 50 states, Hawaii has the 7th highest birth rate, the 4th highest abortion rate, and the 48th highest death rate (only Alaska and Utah are lower). Convert these three simple rankings for Hawaii's birth, abortion, and death rates into percentile rankings.

2. The 70 highest dams in the world have an average height of 206 meters with a standard deviation of 35 meters.
   a. The Hoover, Glen Canyon, and Grand Coulee dams have heights of 221, 216, and 168 meters, respectively. Convert each of these heights into a $z$-score.
   b. Three Russian dams, the Nurek, Toktogul, and Charvak have heights with $z$-scores of +2.69, +0.28, and −1.09, respectively. Convert each of these $z$-scores into a raw score.

3. According to *Consumer Reports* magazine, the costs per pound of protein for 20 major-brand beef hot dogs are as follows: $14.23, $21.70, $14.49, $20.49, $14.47, $15.45, $25.25, $24.02, $18.86, $18.86, $30.65, $25.62, $8.12, $12.74, $14.21, $13.39, $22.31, $19.95, $22.90, and $19.78. If the least expensive is considered the top ranked, what is the position of Thorn Apple Valley Brand beef hot dogs at $14.23 per pound of protein? Give the answer as a simple ranking, as a percentile ranking, and as a $z$-score.

4. The average cost per ounce for glass cleaners is 7.7¢ with a standard deviation of 2.5¢.
   a. Make a diagram with one horizontal line showing the $z$-scores −3, −2, −1, 0, 1, 2, 3, and a second line showing the corresponding raw scores.
   b. What is the $z$-score of Windex with a cost of 10.1¢?

5. The average number of degree-days during the month of January in Alaskan cities is 1751 with a standard deviation of 552.
   a. Make a diagram showing the $z$-scores −3, −2, −1, 0, 1, 2, 3, and the corresponding raw scores.
   b. What is the $z$-score of Fairbanks with 2319 degree-days?

6. In a certain southwestern city the air pollution index averages 62.5 during the year with a standard deviation of 18.0. Assume that the empirical rule is appropriate.
   a. The index will fall within what interval 95% of the time?
   b. For what percentage of the time will the index fall below 26.5?

7. During a recent year the average daily price of gold was $475 with a standard deviation of $55.
   a. At least 75% of the daily prices were within what interval?
   b. At least what percentage of the daily prices were between $310 and $640?

## Study Questions

1. What does Chebyshev's theorem say about the percentage of values that are three or more standard deviations from the mean? About the percentage of values that are more than four standard deviations from the mean?

2. Consider the sample from Exercise 3.
   a. What percentage of this sample fall within one standard deviation of the mean? Within two standard deviations? Within three?
   b. Compare these answers to the results predicted by Chebyshev's theorem and predicted by the empirical rule. Do the results suggest bell-shaped data?

# 4
# SHAPE

There are a variety of ways to organize and arrange data. Much information can be put into tables, but these arrays of bare figures tend to be spiritless and sometimes even forbidding. Some form of graphical display is often best for seeing patterns and shapes and for giving an immediate impression of everything about the data.

In this chapter we develop an understanding of the important visual representation of data called a *histogram*, and then briefly introduce two other displays: *stem and leaf* and *box and whisker*.

## HISTOGRAMS

**Example 4.1**    Suppose there are 2000 families in a small town and the distribution of children among them is as follows: 300 families are childless, 400 have one child, 700 have two children, 300 have three, 100 have four, 100 have five, and 100 have six. These data can be displayed in the following *bar graph*:

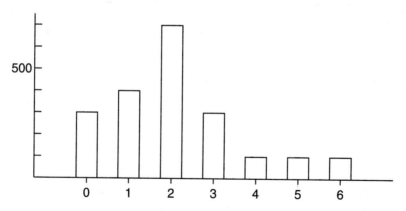

The frequencies of different results are indicated by the *heights* of the bars representing these results.

A *histogram* can be constructed from the above bar graph by widening each bar until the sides meet at a point halfway between.

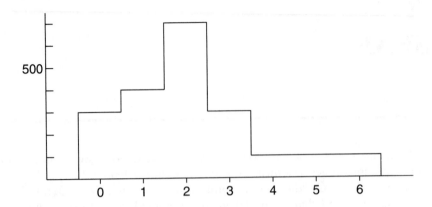

Sometimes, instead of labeling the vertical axis with *frequencies*, it is more convenient or more meaningful to use *relative frequencies*, that is, frequencies divided by the total number in the population.

| Number of Children | Frequency | Relative Frequency |
|:---:|:---:|:---:|
| 0 | 300 | 300/2000 = .150 |
| 1 | 400 | 400/2000 = .200 |
| 2 | 700 | 700/2000 = .350 |
| 3 | 300 | 300/2000 = .150 |
| 4 | 100 | 100/2000 = .050 |
| 5 | 100 | 100/2000 = .050 |
| 6 | 100 | 100/2000 = .050 |

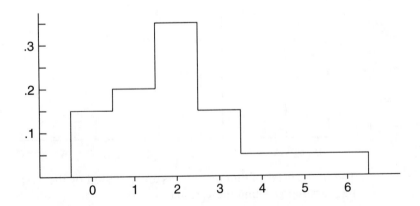

Note that the shape of the histogram is the same whether the vertical axis is labeled with frequencies or with relative frequencies. Sometimes we show both frequencies and relative frequencies on the same graph.

**Example 4.2**    Consider the following histogram displaying 40 salaries paid to the top level executives of a large company.

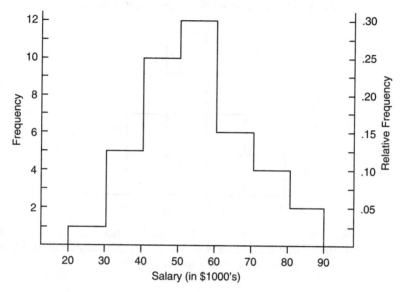

Salary (in $1000's)

What can we learn from this histogram?

For example, none of the executives earned more than $90,000 or less than $20,000. Twelve earned between $50,000 and $60,000. Twenty-five percent earned between $40,000 and $50,000. Note how this histogram shows the number of items (here, salaries) falling *between* certain values, whereas the preceding histogram showed the number of items (families) falling *at* each value.

**Example 4.3**    Consider the following histogram where the vertical axis has not been labeled. What can we learn from this histogram?

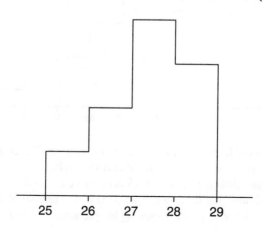

***Answer:*** It is impossible to determine the actual frequencies; however, we can determine the relative frequencies by noting the fraction of the total *area* that is over any interval:

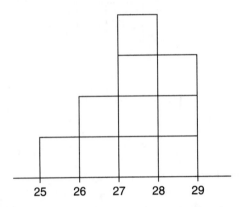

We can divide the area into ten equal portions, and then note that 1/10 or 10% of the area is above 25–26, 20% is above 26–27, 40% is above 27–28, and 30% is above 28–29.

Although it is usually not possible to divide histograms so nicely into ten equal areas, the principle of relative frequencies corresponding to relative areas will still apply.

**Example 4.4**    The following histogram indicates the relative frequencies of ages of U.S. scientists in 1967. What can we learn from this histogram?

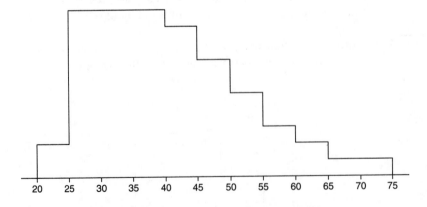

***Answer:*** If we compare areas (or count small rectangles!), we can conclude that 50% of the scientists are between 25 and 40 years of age, 20% are between 45 and 55 years, and so on. Now if we also knew that there were 300,000 U.S. scientists in 1967, we could convert these percentages to frequencies, such as 150,000 scientists between 25 and 40 years of age, or 60,000 scientists between 45 and 55 years of age.

# HISTOGRAMS AND MEASURES OF CENTRAL TENDENCY

Suppose we have a detailed histogram such as this one:

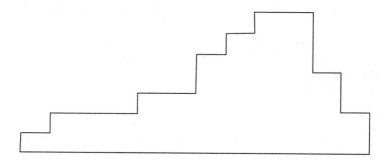

Our measures of central tendency fit naturally into such a diagram.

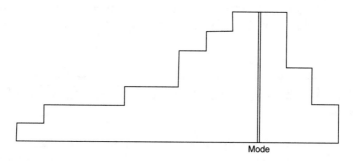

Mode

The *mode* is defined as the most frequent value, so it is the point or interval at which the graph is highest.

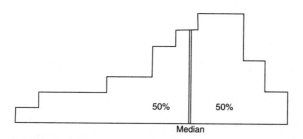

50%    50%

Median

The *median* divides a distribution in half, so it is represented by a line that divides the area of the histogram in half.

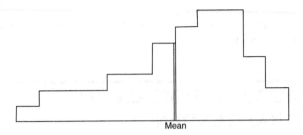

Mean

The *mean* is affected by the spacing of all the values. Therefore, if the histogram is considered to be a solid region, the mean corresponds to a line passing through the center of gravity.

The above distribution, spread thinly far to the low side, is said to be *skewed to the left*. Note that in this case the mean is less than the median. Similarly, a distribution spread far to the high side is *skewed to the right*, and its mean is greater than its median.

**Example 4.5**   Suppose that the faculty salaries at a college have a median of $32,500 and a mean of $38,700. What does this indicate about the shape of the distribution of the salaries?

*Answer:* The median is less than the mean, so the salaries are probably skewed to the right. There are a few highly paid professors with the bulk of the faculty at the lower end of the pay scale.

# HISTOGRAMS, *Z*-SCORES, AND PERCENTILE RANKINGS

We have seen that relative frequencies are represented by relative *areas* and so labeling the vertical axis is not crucial. If we know the standard deviation, the horizontal axis can be labeled in terms of *z-scores*. In fact, if we are given the percentile rankings of various *z*-scores, we can construct a histogram.

**Example 4.6**   Suppose we are given these data:

| z-score: | $-2$ | $-1$ | 0 | 1 | 2 |
|---|---|---|---|---|---|
| Percentile ranking: | 0 | 20 | 60 | 70 | 100 |

and asked to construct a histogram.

We note that the entire area is less than *z*-score $+2$ and greater than *z*-score $-2$. Also, 20% of the area is between *z*-scores $-2$ and $-1$, 40% is between $-1$ and 0, 10% is between 0 and 1, and 30% is between 1 and 2. Thus the histogram is as follows:

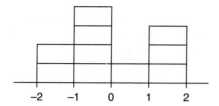 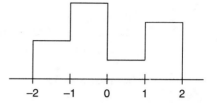

Now suppose that we are given four in-between $z$-scores as well:

| z-score | Percentile ranking |
|---------|-------------------|
| 2.0 | 100 |
| 1.5 | 80 |
| 1.0 | 70 |
| 0.5 | 65 |
| 0.0 | 60 |
| −0.5 | 30 |
| −1.0 | 20 |
| −1.5 | 5 |
| −2.0 | 0 |

Then we have this histogram:

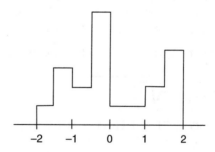

With a 1000 $z$-scores perhaps the histogram would look like this:

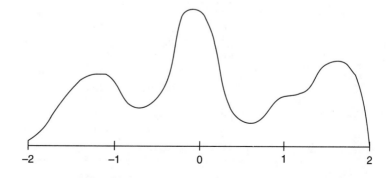

The height at any point is meaningless; what is important is relative *areas*. For example, in the final diagram above, what percent of the area is between z-scores of +1 and +2?

***Answer:*** Still 30%.

What percent is to the left of 0?

***Answer:*** Still 60%.

It is often useful to plot two horizontal scales, one with raw scores and one with the corresponding z-scores.

**Example 4.7**   The heights (in feet) of the 71 mountains over 10,000 feet tall are as follows: 29,028; 28,250; 28,208; 26,810; 26,660; 26,504; 25,645; 25,447; 25,355; 25,230; 24,902; 24,757; 24,590; 24,406; 23,997; 23,891; 22,831; 22,572; 22,310; 22,211; 22,205; 22,057; 21,489; 21,276; 20,561; 20,320; 19,850; 19,347; 19,340; 19,199; 18,855; 18,481; 18,376; 18,008; 17,999; 17,887; 17,058; 17,054; 16,946; 16,864; 16,795; 16,558; 16,503; 15,771; 15,584; 15,203; 15,157; 14,688; 14,494; 14,433; 14,410; 14,272; 14,162; 14,110; 14,022; 13,845; 13,796; 13,680; 13,665; 13,642; 13,350; 13,323; 12,972; 12,461; 12,389; 12,349; 12,172; 12,060; 11,411; 11,053; and 10,457.
Straightforward arithmetic yields $\mu = \dfrac{\Sigma x}{n} = \dfrac{1315563}{71} = 18529$

and $n\sigma = \sqrt{\dfrac{\Sigma x^2}{n} - \mu^2} = \sqrt{25,084,099} = 5008$.   Counting elements (1/71 = 1.41% per element) yields percentile rankings.

| z-score | Raw-score | Number less | Percentile ranking |
|---------|-----------|-------------|--------------------|
| 2.5 | 31049 | 71 | 100 |
| 2.0 | 28545 | 70 | 98.6 |
| 1.5 | 26041 | 65 | 91.5 |
| 1.0 | 23537 | 55 | 77.5 |
| 0.5 | 21033 | 47 | 66.2 |
| 0.0 | 18529 | 40 | 56.3 |
| −0.5 | 16025 | 28 | 39.4 |
| −1.0 | 13521 | 11 | 15.5 |
| −1.5 | 11017 | 1 | 1.4 |
| −2.0 | 8513 | 0 | 0 |

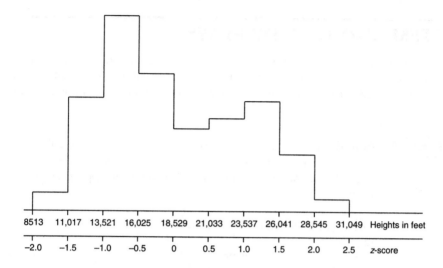

8513   11,017   13,521   16,025   18,529   21,033   23,537   26,041   28,545   31,049   Heights in feet

−2.0   −1.5   −1.0   −0.5   0   0.5   1.0   1.5   2.0   2.5   z-score

**Example 4.8**   Consider the following murder rates (per 100,000 people) for each of the 50 states: AL, 13.3; AK, 12.9; AZ, 9.4; AR, 9.1; CA, 11.7; CO, 7.3; CT, 4.2; DE, 6.7; FL, 11.0; GA, 14.4; HI, 6.7; ID, 5.4; IL, 9.9; IN, 6.2; IA, 2.6; KS, 5.7; KY, 9.0; LA, 15.8; ME, 2.7; MD, 8.2; MA, 3.7; MI, 10.6; MN, 2.0; MS, 12.6; MO, 10.4; MT, 4.8; NE, 3.0; NV, 15.5; NH, 1.4; NM, 10.2; NY, 10.3; NC, 10.8; ND, 1.2; OH, 6.9; OK, 8.5; OR, 5.0; PA, 6.2; RI, 4.0; SC, 11.5; SD, 1.9; TN, 9.4; TX, 14.2; UT, 3.7; VT, 3.3; VA, 8.8; WA, 4.6; WV, 6.8; WI, 2.5; WY, 7.1.

Using our basic formulas gives $\mu = 7.57$ and $\sigma = 3.91$. We calculate raw scores corresponding to various $z$-scores by using $x = 7.57 + 3.91z$. Finally, we count elements (2% per element) to obtain percentile rankings.

| z-score: | −2 | −1.5 | −1 | −0.5 | 0 | 1 | 1.5 | 2 | 2.5 | 3 |
|---|---|---|---|---|---|---|---|---|---|---|
| raw-score: | −0.25 | 1.71 | 3.66 | 5.62 | 7.57 | 9.53 | 11.48 | 13.43 | 15.39 | 17.34 |
| % ranking: | 0 | 4 | 18 | 36 | 54 | 68 | 82 | 92 | 96 | 100 |

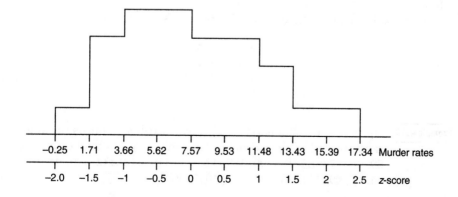

−0.25   1.71   3.66   5.62   7.57   9.53   11.48   13.43   15.39   17.34   Murder rates

−2.0   −1.5   −1   −0.5   0   0.5   1   1.5   2   2.5   z-score

# STEM AND LEAF DISPLAYS

Although a histogram may show how many scores fall into each grouping or interval, the exact values of individual scores are often lost. An alternative pictorial display, called a *stem and leaf display*, retains this individual information.

**Example 4.9**   Consider the set {25, 33, 28, 31, 45, 52, 37, 31, 46, 33, 20}. Let 2, 3, 4, and 5 be place holders for 20, 30, 40, and 50. List the last digit of each value from the original set after the appropriate place holder.

The result is the *stem and leaf display* of these data:

| Stems | Leaves |
|:-----:|:-------|
| 2 | 5 8 0 |
| 3 | 3 1 7 1 3 |
| 4 | 5 6 |
| 5 | 2 |

Drawing a continuous line around the leaves results in a horizontal histogram:

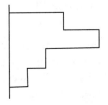

Note that the stem and leaf display gives the shape of the histogram and, unlike the histogram, indicates the values of the original data.

A *revised stem and leaf display* is obtained by rearranging the numbers in each row in ascending order. This ordered display shows a second level of information from the original stem and leaf picture.

The revised display of the data in Example 4.9 is as follows:

| | |
|:-----:|:-------|
| 2 | 0 5 8 |
| 3 | 1 1 3 3 7 |
| 4 | 5 6 |
| 5 | 2 |

**Example 4.10**   Suppose the distribution of 25 advertised house prices (in thousands of dollars) in a certain community is given by the set: {56, 89, 165, 73, 83, 145, 90, 189, 127, 77, 110, 112, 132, 120, 94, 130, 84, 65, 99, 154, 86, 120, 122, 103, 130}.

One possible stem and leaf display of these data is as follow:

| 50–74 | 56 65 73 |
|---|---|
| 75–99 | 77 83 84 86 89 90 94 99 |
| 100–124 | 03 10 12 20 20 |
| 125–149 | 22 27 30 30 32 45 |
| 150–174 | 54 65 |
| 175–200 | 89 |

Note that the stems were chosen to be intervals of length 25, and that the "1" of 100 is left out of the leaves so that they align vertically.

# BOX AND WHISKER DISPLAYS

A *box and whisker display* is a visual representation of dispersion that shows the largest value, the smallest value, the median, the median of the top half of the set, and the median of the bottom half of the set.

**Example 4.11**  The total farm product indexes for the years from 1919 through 1945 (with 1910–14 as 100) are as follows: 215, 210, 130, 140, 150, 150, 160, 150, 140, 150, 150, 125, 85, 70, 75, 90, 115, 120, 125, 100, 95, 100, 130, 160, 200, 200, 210. (Note the instability of prices received by farmers!) The largest value is 215, the smallest is 70, the median is 140, the median of the top half is 160, and the median of the bottom half is 100. A *box and whisker display* of these five numbers is as follows:

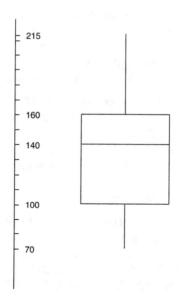

Note that the display consists of two "boxes" together with two "whiskers"— hence the name. The boxes show the spread of the two middle quarters; the whiskers, the spread of the two outer quarters. This relatively simple display conveys information not immediately available from histograms or stem and leaf displays.

# EXERCISES

In Exercises 1–4, draw an appropriate histogram for each set of data.

1. The 54 major brands of cottage cheese on the market have the following numbers of calories per 4-ounce serving:

| Calories | 75–84 | 85–94 | 95–104 | 105–114 | 115–124 |
|---|---|---|---|---|---|
| Number of brands | 15 | 10 | 9 | 15 | 5 |

2. The 60 longest rivers in the world have lengths distributed as follows:

| Length (miles) | 1000–1499 | 1500–1999 | 2000–2499 | 2500–2999 | 3000–3499 | 3500–3999 | 4000–4499 |
|---|---|---|---|---|---|---|---|
| Number of rivers | 21 | 22 | 4 | 8 | 2 | 2 | 1 |

(The Nile is the longest with a length of 4145 miles, and the Amazon is second at 3900 miles.)

3. Use the farm product index data in Example 4.11 for this exercise. Plot two horizontal scales, one with raw scores and one with corresponding $z$-scores. Use intervals of 1.0 along the $z$-score scale for the histogram.

4. The ages of U.S. presidents at inauguration were as follows: 57, 61, 57, 57, 58, 57, 61, 54, 68, 51, 49, 64, 50, 48, 65, 52, 56, 46, 54, 49, 50, 47, 55, 55, 54, 42, 51, 56, 55, 51, 54, 51, 60, 62, 43, 55, 56, 61, 52, 68, 65, and 46. Plot two horizontal scales, one with raw scores and one with corresponding $z$-scores. Use intervals of 0.5 along the $z$-score scale for the histogram.

5. a. Prepare a stem and leaf display for the set {126, 195, 149, 122, 189, 164, 228, 177, 165, 150, 169, 127, 176, 147, 148, 159, 128, 122, 150, 193, 207, 164, 168, 110, 155, 127, 152, 174, 190, 219, 125, 193, 141, 127, 155, 133, 150, 162, 168, 128, 125, 137, 146, 120, 154, 176, 166, 117, 154, 137}, which represents the 1988 per capita personal income (in hundreds of dollars) for the 50 states.

   b. Draw a box and whisker display of the above data.

6. According to the *1992 NAEP Trial State Assessment* the average mathematics proficiency scores in eighth grade for 41 states were as follows: AL, 251; AZ, 265; AR, 255; CA, 260; CO, 272; CT, 273; DE, 262; FL, 259; GA, 259; HI, 257; ID, 274; IN, 269; IA, 283; KY, 261; LA, 249; ME, 278; MD, 264; MA, 272; MI, 267; MN, 282; MS, 246; MO, 270; NE, 277; NH, 278; NJ, 271;

NM, 259; NY, 266; NC, 258; ND, 283; OH, 267; OK, 267; PA, 271; RI, 265; SC, 260; TN, 258; TX, 264; UT, 274; VA, 267; WV, 258; WI, 277; WY, 274.

a. Draw a box and whisker display.
b. Prepare a stem and leaf display.
c. Find the mean and standard deviation of the 41 scores.
d. According to Chebyshev's theorem, how many of the scores should be within two standard deviations of the mean? How many of the scores are actually within this range?
e. Do you think the empirical rule applies? According to the empirical rule, how many of the scores should be within one standard deviation of the mean? Within two? Within three? How many of the scores are actually within one, two, and three standard deviations of the mean?

7. A 1995 poll by the Program for International Policy asked respondents what percentage of the U.S. budget they thought went to foreign aid. The mean response was 18%, and the median was 15%. (The actual amount is less than 1%.) What do these responses indicate about the shape of the distribution of all the responses?

## Study Questions

1. Consider the following two histograms:

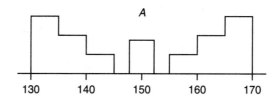

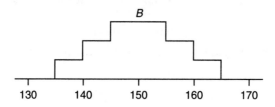

True or false?
a. Both sets have about the same mean.
b. The variance of set *A* is greater than the variance of set *B*.
c. Chebyshev's theorem applies to both sets.
d. The empirical rule applies only to set *A*.
e. You can be sure that the standard deviation of set *A* is greater than 5.

2. Consider the following histogram:

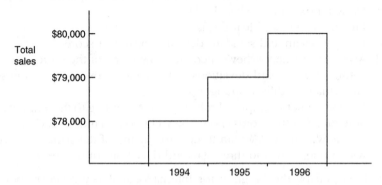

a. What is misleading about the vertical axis and the resulting sales picture?
b. Draw a histogram showing the same information, this time, but with a vertical axis starting at zero.

3. Consider the following histogram:

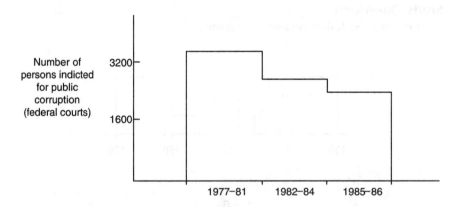

Discuss what is misleading about this histogram.

4. In your daily newspaper find a histogram that misrepresents the information given.

# Part 2

# PROBABILITY

---

*In the world around us, unlikely events sometimes take place. At other times, events that seem inevitable do not occur. Because of the myriad and minute origins of various happenings, it is often impracticable, or simply impossible, to predict exact outcomes. However, while we may not be able to foretell a specific result, we can sometimes assign what is called a* probability *to indicate the likelihood that a particular event will occur.*

*For our study of statistics, we need an understanding of the probability that a given elementary event will happen many times, each time under the same circumstances. We want to be able to deduce the chance or prospect of occurrence of such events, and then use the result to draw inferences about more complex circumstances for which complete information is unavailable. For example, we might analyze the past movements of various stock prices given various economic conditions, calculate specific probabilities, and then ask what can be said, with what degree of confidence, about future movements.*

*In Chapters 5–7 we concentrate on the development of the specific techniques necessary to appreciate the basic principles of statistical analysis which will be considered later. In particular, we need to understand (1) how to use counting techniques; (2) how to calculate elementary probabilities, including* binomial *probabilities, and (3) how to calculate* expected *values.*

# 5
# COUNTING

Counting techniques are important in solving probability problems. In this chapter we develop the concept of *combinations*, a counting procedure critical in later applications. Along the way we learn the *multiplicative rule* and the way to count *permutations*.

Examples 5.1 and 5.2 illustrate the use of the multiplicative rule.

**Example 5.1**    A company must choose one of eight applicants for a secretarial position, and one of six applicants for a janitorial position. In how many ways can this be done?

*Answer:* We can pair up each of the eight secretarial applicants with each of the six janitorial applicants for a total of $8 \times 6 = 48$ ways.

| 8 | 6 |
|---|---|
| ways of picking secretaries | ways of picking janitors |

$8 \times 6 = 48$

**Example 5.2**    A manufacturer must choose one of two warehouses, one of five assembly plants and one of three distribution centers. In how many ways can this be done?

*Answer:* There are $2 \times 5 = 10$ ways of picking a warehouse and an assembly plant, and for each of these there are 3 ways to pick a distribution center, for a total of $10 \times 3 = 30$ ways.

| 2 | 5 | 3 |
|---|---|---|
| ways of picking warehouses | ways of picking assembly plants | ways of picking distribution centers |

$2 \times 5 \times 3 = 30$

The multiplicative rule can be stated in more general terms.

## MULTIPLICATIVE RULE

If one event can occur in *m* ways, and, for each of these ways, a second event can occur in *n* ways, then the two events can occur together in *mn* ways. The same principle can be extended to three or more events, as in Example 5.2.

The multiplicative rule is the underlying principle behind various counting techniques.

**Example 5.3**   A president and vice president are to be picked from six candidates. In how many ways can this be done?

*Answer:* There are 6 choices for president and then 5 remaining choices for vice president for a total of 6 × 5 = 30 ways.

| 6 | 5 |
|---|---|

6 x 5 = 30

ways of choosing president  ways of choosing vice-president

**Example 5.4**   Three bonuses, one for $10,000, one for $5000 and one for $2000, are to be given to three employees chosen from a group of ten. In how many ways can this be done?

*Answer:* There are 10 choices for the $10,000 bonus, 9 remaining choices for the $5000 bonus, and then 8 remaining choices for the $2000 bonus for a total of 10 × 9 × 8 = 720 ways.

| 10 | 9 | 8 |
|----|---|---|

10 x 9 x 8 = 720

choices for $10,000 bonus   choices for $5000 bonus   choices for $2000 bonus

The answers to Examples 5.3 and 5.4 can be expressed in terms of *factorials* (3! = 3 × 2 × 1 = 6, 5! = 5 × 4 × 3 × 2 × 1 = 120, etc.) as follows:

$$6 \times 5 = \frac{6 \times 5 \times 4 \times 3 \times 2 \times 1}{4 \times 3 \times 2 \times 1} = \frac{6!}{4!}$$

$$10 \times 9 \times 8 = \frac{10 \times 9 \times 8 \times 7 \times 6 \times 5 \times 4 \times 3 \times 2 \times 1}{7 \times 6 \times 5 \times 4 \times 3 \times 2 \times 1} = \frac{10!}{7!}$$

Note that 6 × 5 has 2 factors, 6 − 2 = 4, and 6 × 5 = 6!/(6 − 2)!. Also 10 × 9 × 8 has 3 factors, 10 − 3 = 7, and 10 × 9 × 8 = 10!/(10 − 3)!.

Similarly, if we have *r* factors beginning with *n*, we get $n!/(n - r)!$. This is the basis of the permutation rule.

## PERMUTATION RULE

The number of ways of choosing *r* distinct objects from among *n* objects, where "order is important" is $n!/(n - r)!$. Each of the different ways of making this choice is called a *permutation*, and the total number of such permutations is denoted by $P(n,r)$. Thus, $P(n,r) = n!/(n - r)!$. We can see that "order is important" in Examples 5.3 and 5.4; for example, choosing Tom to be president and Mary to be vice president is different from choosing Mary to be president and Tom to be vice president.

**Example 5.5**  For a special advertising campaign, a company plans to pick one of its nine products for TV, one for radio, one for magazines, and one for newspapers. In how many ways can this be done?

$$P(9,4) = \frac{9!}{(9 - 4)!} = \frac{9!}{5!} = 3024$$

In situations where order is *not* important, we must modify the counting procedure.

**Example 5.6**  In how many ways can a two person committee be chosen from six people?

*Answer:* Compare this problem with Example 5.3. In that case, the answer was $6 \times 5 = 30$, but that answer is too large for this example. While choosing Tom as president and Mary as vice president is different from choosing Mary as president and Tom as vice president, there is only *one* two-person committee consisting of Tom and Mary. Thus 30 is two times too large, and the correct answer is $30/2 = 15$.

**Example 5.7**  In how many ways can three employees be chosen from a group of ten to receive identical $4000 bonuses?

*Answer:* Compare this problem with Example 5.4. In that case, the answer was $10 \times 9 \times 8 = 720$, but again, that answer is too large for this example. There is only *one* way to give Jane, David, and Ann the same $4000 raise, but how many ways can we distribute three different raises among three people? The answer is $3 \times 2 \times 1$ because there are three choices for the big raise, two remaining choices for the middle raise, and only one remaining choice for whom to give the last raise. Thus 720 is $3 \times 2 \times 1 =$

3 ! = 6 times too large for this example, and the correct answer is 720/6 = 120.

The answers to Examples 5.6 and 5.7 can be expressed in terms of factorials as follows: $6 \times 5$ has 2 factors, and we had to divide $6 \times 5 = 6!/4!$ by the number 2 or 2!. $10 \times 9 \times 8$ has 3 factors, and we had to divide $10 \times 9 \times 8 = 10!/7!$ by the number 6 or 3!. Also, $6!/4!$ divided by 2! is $\dfrac{6!}{4!\,2!}$ and $10!/7!$ divided by 3! is $\dfrac{10!}{7!\,3!}$. Similarly, if we divide $n!/(n-r)!$ by the number $r!$, the result is $\dfrac{n!}{(n-r)!\,r!}$.

# COMBINATION RULE

The number of ways of choosing $r$ distinct objects from $n$ objects, where "order is not important," is $n!/(n-r)!\,r!$. Each of the different ways of making this choice is called a *combination*, and the total number of such combinations is denoted by $C(n,r)$ or $\binom{n}{r}$. Thus

$$C(n,r) = \binom{n}{r} = \frac{n!}{(n-r)!\,r!}$$

**Example 5.8**    In how many ways can a company select four of its nine products for an advertising campaign? (Note the difference from Example 5.5.)

**Answer:**

$$C(9,4) = \binom{9}{4} = \frac{9!}{(9-4)!\,4!} = \frac{9!}{5!4!} = 126$$

For ease in future calculations, we note how to simplify factorial divisions. For instance,

$$\frac{10!}{6!} = \frac{10 \times 9 \times 8 \times 7 \times 6!}{6!} = 10 \times 9 \times 8 \times 7 = 5040$$

$$\frac{8!}{3!} = \frac{8 \times 7 \times 6 \times 5 \times 4 \times 3!}{3!} = 8 \times 7 \times 6 \times 5 \times 4 = 6720$$

When evaluating combinations, it is simplest to cancel the larger factorial in the denominator against part of the factorial in the numerator.

**Example 5.9**

$$C(7,4) = \frac{7!}{3!4!} = \frac{7 \times 6 \times 5}{3 \times 2} = 35$$

We canceled the $4!$ in the denominator against part of the $7!$ in the numerator, and we were left with $3! = 3 \times 2$ in the denominator. Similarly:

$$C(12,10) = \frac{12!}{2!10!} = \frac{12 \times 11}{2} = 66$$

$$C(20,4) = \frac{20!}{16!4!} = \frac{20 \times 19 \times 18 \times 17}{4 \times 3 \times 2} = 4845$$

We define $0!$ to be 1, and so, for example, $C(5,5) = \frac{5!}{0!5!} = 1$. This result is expected since there is only one way of choosing five items from a group of five when order is not important. Similarly, $\binom{4}{4} = 1$, $\binom{2}{2} = 1$, $C(10,10) = 1$, etc. Another combination in which $0!$ arises has the form $C(6,0) = \frac{6!}{6!0!} = 1$. (There is exactly one way to choose no item from a group of six.)

Examples of other readily computed combinations are $C(57,1) = \frac{57!}{56!1!} = 57$ (there are 57 ways to pick one item from a group of 57) and $C(32,31) = 32$ (there are 32 ways to pick all but one item from a group of 32). Similarly, $C(85,1) = 85$ and $C(23,22) = 23$.

The results of other combinations are not as obvious and require some work. However, patterns can be noticed. For example, consider the following:

$$\binom{0}{0} = 1$$

$$\binom{1}{0} = 1 \quad \binom{1}{1} = 1$$

$$\binom{2}{0} = 1 \quad \binom{2}{1} = 2 \quad \binom{2}{2} = 1$$

$$\binom{3}{0} = 1 \quad \binom{3}{1} = 3 \quad \binom{3}{2} = 3 \quad \binom{3}{3} = 1$$

$$\binom{4}{0} = 1 \quad \binom{4}{1} = 4 \quad \binom{4}{2} = 6 \quad \binom{4}{3} = 4 \quad \binom{4}{4} = 1$$

$$\binom{5}{0} = 1 \quad \binom{5}{1} = 5 \quad \binom{5}{2} = 10 \quad \binom{5}{3} = 10 \quad \binom{5}{4} = 5 \quad \binom{5}{5} = 1$$

These numbers form a pattern known as *Pascal's triangle*:

```
            1
          1   1
        1   2   1
      1   3   3   1
    1   4   6   4   1
  1   5  10  10   5   1
  ...        ...        ...
```

If one line of Pascal's triangle is known, how can the next line be found?

*Answer:* The next line will start and end with 1. Each other number can be found by adding the two numbers immediately above and slightly to the right and left. For example, the next line after the last one above is

1    1+5=6    5+10=15    10+10=20    10+5=15    5+1=6    1

Thus,

$$\binom{6}{0} = 1 \quad \binom{6}{1} = 6 \quad \binom{6}{2} = 15 \quad \binom{6}{3} = 20 \quad \binom{6}{4} = 15 \quad \binom{6}{5} = 6 \quad \binom{6}{6} = 1$$

Finally, for later use, we comment on how to find $\binom{n}{r}$ if we know $\binom{n}{r-1}$.

$$\frac{n-r+1}{r}\binom{n}{r-1} = \binom{n}{r} \qquad \text{(See Study Question 1)}$$

**Example 5.10**  Given that $\binom{8}{3} = 56$, what is $\binom{8}{4}$?

*Answer:*

$$\frac{8-4+1}{4} = \frac{5}{4} \qquad and \qquad \frac{5}{4} \times 56 = 70$$

Given that $\binom{15}{9} = 5005$, what is $\binom{15}{10}$?

*Answer:*

$$\frac{15-10+1}{10} = \frac{6}{10} \qquad and \qquad \frac{6}{10} \times 5005 = 3003$$

# EXERCISES

1. In his novel *Alaska*, James Michener talks about the different factors that must combine to produce an ice age. Then, mathematically, he explains that "if four different factors in an intricate problem operate in cycles of 13, 17, 23, and 37 years respectively, and if all have to coincide to produce the desired result, you might have to wait 188,071 years ...." Explain how Michener calculated 188,071.

2. As reported in the Allentown *Morning Call*, Boston Chicken advertised that their customers had a choice of 3360 combinations of three side dishes from among their 16 side dishes offered. A high school teacher pointed out the mistake; can you?

3. Johann Gregor Mendel characterized garden peas by the following seven

attributes: the stems are either long or short; the flowers are either red or white, and are either terminal or axial; the pods are either green or yellow, and are either constricted or inflated; and the seeds are either green or yellow, and are either smooth or wrinkled. How many different varieties of garden peas are possible?

## Study Questions

1. Show that $\dfrac{n - r + 1}{r} \dbinom{n}{r - 1} = \dbinom{n}{r}$. [*Hint*: Write out the expressions for $\dbinom{n}{r - 1}$ and for $\dbinom{n}{r}$, and note that $r(r - 1)! = r!$ and $(n - r + 1)(n - r)! = (n - r + 1)!$.]

2. Prove the formula $\dbinom{n}{r} = \dbinom{n - 1}{r} + \dbinom{n - 1}{r - 1}$ which we used in Example 5.10 to determine an entry in Pascal's triangle from the two *above* entries. [Hint: Write out the algebraic expressions for each of the two combinations on the right, and then convert to the common denominator: $r!(n - r)!$.]

3. Suppose we have a set with $n$ elements. Then the number of subsets that contain exactly $r$ elements is $C(n,r)$. What proportion of these subsets contain some designated element of the original set? [*Hint*: Reason that the answer is $C(n - 1, r - 1)/C(n,r)$, and show algebraically that this is equal to $r/n$.]

4. a. What is the total number of subsets of a set with two elements? Of a set with three elements?

   b. Do you think that a set could have exactly ten subsets? Sixteen subsets?

# 6
# PROBABILITY CONCEPTS

The *probability* of a particular outcome of an experiment is a mathematical statement about the likelihood of that event occurring. Probabilities are always between 0 and 1 with a probability close to 0 meaning that an event is unlikely to occur, and a probability close to 1 meaning that the event is likely to occur. The sum of the probabilities of all the separate outcomes of an experiment is always 1. In this chapter we list the basic probability concepts that are needed for our discussion of probability distributions.

## ELEMENTARY PROBABILITY

### Complementary events
The probability that an event will not occur, that is, the probability of its complement, is equal to 1 minus the probability that the event will occur:

$$P(X') = 1 - P(X)$$

**Example 6.1**  If the probability that a company will win a contract is .3, what is the probability that it will not win the contract?

*Answer:* $1 - .3 = .7$.

### Addition principle
If two events are mutually exclusive, that is, they cannot occur simultaneously, then the probability that at least one event will occur is equal to the sum of the respective probabilities of the two events. That is:

$$\text{If } P(X \cap Y) = 0, \text{ then } P(X \cup Y) = P(X) + P(Y)$$

where $X \cap Y$, read "$X$ intersect $Y$," means that both X and Y occur, while $X \cup Y$, read "$X$ union $Y$," means that either $X$ or $Y$ or both occur.

**Example 6.2**  If the probabilities that Jane, Tom, and Mary will be chosen chairperson of the board are .5, .3, and .2, respectively, then the probability that the chairperson will be either Jane or Mary is $.5 + .2 = .7$.

**51**

When two events are not mutually exclusive, then the sum of their probabilities counts their shared occurrence twice. This leads to the

## General addition formula

For any pair of events $X$ and $Y$,

$$P(X \cup Y) = P(X) + P(Y) - P(X \cap Y)$$

**Example 6.3**    Suppose that the probability that a construction company will be awarded a certain contract is .25, the probability that it will be awarded a second contract is .21, and the probability that it will get both contracts is .13. What is the probability that the company will win at least one of the two contracts?

*Answer:* .25 + .21 − .13 = .33

## Independence principle

If the chance that one event will happen is not influenced by whether or not a second event happens, the probability that *both* events will occur is the product of their separate probabilities.

**Example 6.4**    The probability that a student will receive a state grant is 1/3, while the probability that she will be awarded a federal grant is 1/2. If whether or not she receives one grant is not influenced by whether or not she receives the other, what is the probability of her receiving both grants?

*Answer:* $1/3 \times 1/2 = 1/6$

The independence principle can be extended to more than two events; that is, given a sequence of independent events, the probability that *all* will happen is equal to the product of their individual probabilities.

**Example 6.5**    Suppose a reputed psychic in an ESP experiment has called heads or tails correctly on ten successive tosses of a coin. What is the probability that guessing would yield this perfect score?

*Answer:* $(1/2)(1/2)...(1/2) = (1/2)^{10} = 1/1024$

In many applications, such as coin tossing, there are only *two* possible outcomes. For example, on each toss either the psychic guesses correctly or she guesses incorrectly, either a radio is defective or it is not defective, either the workers will go on strike or they will not walk out, either the manager's salary is above $50,000 or it does not exceed $50,000. In some applications such two-outcome situations are repeated many times. For example, we may ask about the chances that at most one of four tires is defective, or that at least three of five unions will vote to go on strike, or that exactly two of three executives have salaries above $50,000. For applications in which a two-outcome situation is repeated some number of times, and the probability of each of the two outcomes remains the same for each repetition, the resulting calculations involve what are known as *binomial probabilities.*

**Example 6.6**    Suppose the probability that a light bulb is defective is .1. What is the probability that four light bulbs are all defective?

*Answer:* $(.1)(.1)(.1)(.1) = (.1)^4 = .0001$

**Example 6.7**    Again suppose the probability that a light bulb is defective is .1. What is the probability that exactly two of three light bulbs are defective.

*Answer:* We subdivide the problem as follows. The probability that the first two bulbs are defective, and the third is good, is $(.1)(.1)(.9) = .009$. (Note that, if the probability of being defective is .1, the probability of being good is .9.) The probability that the first bulb is good, and the other two are defective, is $(.9)(.1)(.1) = .009$. Finally, the probability that the second bulb is good, and the other two are defective is $(.1)(.9)(.1) = .009$. Summing, we find that the probability that exactly two out of three bulbs are defective is $.009 + .009 + .009 = .027$.

**Example 6.8**    If the probability of a defective light bulb is .1, what is the probability that exactly three out of eight light bulbs are defective?

*Answer:* We can subdivide the problem again. For example, the probability that the first, third, and seventh bulbs are defective, and the rest are good, is

$$(.1)(.9)(.1)(.9)(.9)(.9)(.1)(.9) = (.1)^3(.9)^5 = .00059049$$

The probability that the second, third, and fifth bulbs are defective, and the rest are good, is

$$(.9)(.1)(.1)(.9)(.1)(.9)(.9)(.9) = (.1)^3(.9)^5 = .00059049$$

As can be seen, the probability of any particular arrangement of three defective and five good bulbs is $(.1)^3(.9)^5 = .00059049$. How many such arrangements are there? In other words, in how many ways can we pick three of eight positions for the defective bulbs (the remaining five positions will be for good bulbs)? From Chapter 5 the answer is given by *combinations*:

$$C(8,3) = \frac{8!}{5!3!} = \frac{8 \times 7 \times 6}{3 \times 2} = 56$$

Each of these 56 arrangements has a probability of .00059049. Thus, the probability that exactly three out of eight light bulbs are defective is $56 \times .00059049 = .03306744$:

$$C(8,3)(.1)^3(.9)^5 = \frac{8!}{3!5!}(.1)^3(.9)^5 = .03306744$$

**Example 6.9**    Suppose 30% of the employees in a large factory are smokers. What is the probability that there will be exactly two smokers in a randomly chosen five-person work group?

*Answer:* We reason as follows. The probability that a person smokes is 30% = .3, so the probability that he or she does not smoke is $1 - .3 = .7$. The probability of a particular arrangement of two smokers and three nonsmokers is $(.3)^2(.7)^3 = .03087$. The number of such arrangements is $C(5,2) = 5!/2!3! = 10$. Each such arrangement has probability .03087, so the final answer is $10 \times .03087 = .3087$.

$$C(5,2)(.3)^2(.7)^3 = \frac{5!}{2!3!}(.3)^2(.7)^3 = .3087$$

We can state the general principle as the binomial formula.

# BINOMIAL FORMULA

Suppose an experiment has two possible outcomes, called *success* and *failure*, with the probability of success equal to $p$ and the probability of failure equal to $q$ (of course, $p + q = 1$). Suppose further that the experiment is repeated $n$ times, and the outcome at any particular time has no influence over the outcome at any other time. Then the probability of exactly $x$ successes (and thus $n-x$ failures) is

$$C(n, x)\, p^x\, q^{n-x} = \frac{n!}{x!(n - x)!}\, p^x\, q^{n-x}$$

**Example 6.10**    A manager notes that there is a .125 probability that any employee will arrive late for work. What is the probability that exactly one person in a six-person department will arrive late?

*Answer:* If the probability of being late is .125, then the probability of being on time is $1 - .125 = .875$. If one person of six is late, then $6 - 1 = 5$ will be on time. Thus the desired probability is

$$C(6,1)(.125)^1(.875)^5 = 6(.125)(.875)^5 = .385$$

Many, perhaps most, applications of probability involve such phrases as *at least, at most, less than,* and *more than.* In these cases, solutions involve summing two or more cases.

**Example 6.11**  A manufacturer has the following quality control check at the end of a production line: If at least eight of ten randomly picked articles meet all specifications, the whole shipment is approved. If, in reality, 85% of a particular shipment meet all specifications, what is the probability that the shipment will make it through the control check?

*Answer:* The probability of meeting specifications is .85, so the probability of not meeting specifications must be .15. We want the probability that at least eight of ten articles meet specifications, that is, the probability that exactly eight or exactly nine or exactly ten articles meet specifications. We sum the three binomial probabilities:

| Exactly 8 of 10 meet specifications | Exactly 9 of 10 meet specifications | Exactly 10 of 10 meet specifications |
|---|---|---|

$$C(10,8)(.85)^8(.15)^2 + C(10,9)(.85)^9(.15)^1 + C(10,10)(.85)^{10}(.15)^0$$
$$= \frac{10!}{8!2!}(.85)^8(.15)^2 + 10(.85)^9(.15)^1 + (.85)^{10} = .820$$

**Example 6.12**  For the problem in Example 6.11, what is the probability that a shipment in which only 70% of the articles meet specifications will make it through the control check?

*Answer:*

$$C(10,8)(.7)^8(.3)^2 + C(10,9)(.7)^9(.3)^1 + C(10,10)(.7)^{10}(.3)^0$$
$$= 45(.7)^8(.3)^2 + 10(.7)^9(.3) + (.7)^{10} = .383$$

In some situations it is easier to calculate the probability of the complementary event and subtract this value from 1.

**Example 6.13**  Joe DiMaggio had a career batting average of .325. What was the probability that he would get at least one hit in five official times at bat?

*Answer:* We could sum the probabilities of exactly one hit, two hits, three hits, four hits, and five hits. However, the complement of "at least one hit" is "zero hits." The probability of no hit is

$$C(5,0)(.325)^0(.675)^5 = (.675)^5 = .140,$$

and thus the probability of at least one hit in five times at bat is $1 - .140 = .860$.

**Example 6.14**  A grocery store manager notes that 35% of the customers who buy a particular product make use of a store coupon to receive a discount. If seven people purchase the product, what is the probability that fewer than four will use a coupon?

*Answer:* In this situation, "fewer than four" means zero or one or two or three.

$$C(7,0)(.35)^0(.65)^7 + C(7,1)(.35)^1(.65)^6 + C(7,2)(.35)^2(.65)^5$$
$$+ C(7,3)(.35)^3(.65)^4$$
$$= (.65)^7 + 7(.35)(.65)^6 + 21(.35)^2(.65)^5 + 35(.35)^3(.65)^4 = .800$$

Sometimes we are asked to calculate the probability of each of the possible outcomes (the results should sum to 1).

**Example 6.15**  If the probability of a male birth is .51, what is the probability that a five-child family will have all boys? Exactly four boys? Exactly three boys? Exactly two boys? Exactly one boy? All girls?

*Answer:*

$$
\begin{aligned}
P(5 \text{ boys}) &= C(5,5)(.51)^5(.49)^0 = (.51)^5 &= .0345 \\
P(4 \text{ boys}) &= C(5,4)(.51)^4(.49)^1 = 5(.51)^4(.49) &= .1657 \\
P(3 \text{ boys}) &= C(5,3)(.51)^3(.49)^2 = 10(.51)^3(.49)^2 &= .3185 \\
P(2 \text{ boys}) &= C(5,2)(.51)^2(.49)^3 = 10(.51)^2(.49)^3 &= .3060 \\
P(1 \text{ boy}) &= C(5,1)(.51)^1(.49)^4 = 5(.51)(.49)^4 &= .1470 \\
P(0 \text{ boys}) &= C(5,0)(.51)^0(.49)^5 = (.49)^5 &= \underline{.0283} \\
& & 1.0000
\end{aligned}
$$

A table such as the one in Example 6.15 shows the entire *probability distribution*, which in this case refers to a listing of the probabilities of all outcomes.

For later use we need to know how to calculate the probability of $x$ successes if we know the probability of $x - 1$ successes:

$$P(x\ successes)\ =\ C(n, x)p^x\ q^{n-x}$$

$$P(x - 1\ successes)\ =\ C(n, x - 1)p^{x-1}q^{n-x+1}$$

$$[n - (x - 1) = n - x + 1]$$

Note that $P(x$ successes$)$ has one more $p$ and one less $q$ than does $P(x - 1$ successes$)$, so we should expect a factor $p/q$. Also, from Chapter 5 we have $C(n,x) = [(n - x + 1)/x\ ]C(n,x - 1)$. Combining these results gives

$$P(x\ successes)\ =\ \frac{n - x + 1}{x}\frac{p}{q}\ P(x - 1\ successes)$$

Since $q = 1 - p$, this is often written

$$P(x\ successes)\ =\ \frac{n - x + 1}{x}\frac{p}{1 - p}\ P(x - 1\ successes)$$

**Example 6.16** A sharpshooter can hit a bullseye target 95% of the time. Given that the probability of exactly eight bullseyes in ten shots is .0746, what is the probability of exactly nine bullseyes in ten shots?

*Answer:*

$$P(10\ bullseyes)\ =\ \frac{10 - 9 + 1}{9}\frac{.95}{1 - .95}\ (.0746) = .315$$

# EXERCISES

1. As reported in *The New York Times* (February 19, 1995), the Russian Health Ministry announced that one-quarter of the country's hospitals had no sewage system and one-seventh had no running water. What is the probability that a Russian hospital will have at least one of these problems if:
   a. the two problems are independent?
   b. hospitals with the running water problem are a subset of those with the sewage problem?

2. In the November 27, 1994, issue of *Parade* magazine, the "*Ask Marilyn*" section contained this question:
   "Suppose a person was having two surgeries performed at the same time. If the chances of success for surgery *A* are 85%, and the chances of success for surgery *B* are 90%, what are the chances that both would fail?"
   What do you think of Marilyn's solution: $(.15)(.10) = .015$ or 1.5%?

3. In November 1994, Intel announced that a "subtle flaw" in its Pentium chip would affect one in 9,000,000,000 division problems. Suppose a computer performed 20,000,000 divisions (a not unreasonable number) in the course

of a particular program. What is the probability of no error? Of at least one error?

4. According to a CBS/*New York Times* poll taken in 1992, 15% of the public has responded to a telephone call-in poll. In a random group of five people, what is the probability that exactly two have responded to a call-in poll? That at least two have responded? That at most two have responded?

5. In a 1974 "Dear Abby" letter a woman lamented that she had just given birth to her eighth child, and all were girls! Her doctor had assured her that the chance of the eighth child being a girl was only one in 100.
   a. What was the real probability that the eighth child would be a girl?
   b. Before the birth of the first child, what was the probability that the woman would give birth to eight girls in a row?

6. The yearly mortality rate for American men from prostate cancer has been constant for decades at about 25 of every 100,000 men. (This rate has not changed in spite of new diagnostic techniques and new treatments.) In a group of 100 American men, what is the probability that at least one will die from prostate cancer in a given year?

## Study Questions

1. (*Geometric distribution*) Suppose the probability that someone will make a major mistake on an income tax return is .08. One day, an IRS agent plans to audit as many returns as necessary until she finds one with a major mistake. What is the probability that a major mistake will be found on the first return? The second? The third? The fourth? The fifth? Note that, while the probability of a major mistake on any given return is always .08, the probability of which return the first major mistake is found on steadily decreases. The associated cumulative probability distribution is called a *geometric distribution*. Reason why, in this question, if the first three returns have no major mistake, the probability that more than five additional returns will have to be audited to find a major mistake is the same as the probability at the beginning of the day that more than five returns will have to be audited.

2. (*Negative binomial distribution*) During the 1990 World Series, the Oakland Athletics were heavily favored over the Cincinnati Reds. Suppose the probability that Cincinnati would win any given game was 4/9. What was the probability that Cincinnati would win the series, that is, would win four games before Oakland won four games? [*Hint:* The probabilities of separate cases must be summed; for each case, consider how many wins and losses occurred before the final game.] The above gives rise to what is called a *negative binomial distribution*. For a negative binomial distribution, as well as for the geometric distribution described in Study Question 1, the number of "successes" is fixed, while the total number of trials is allowed to vary. However, for the original binomial distribution, the number of "successes" is allowed to vary, while the number of trials is fixed.

# 7
# EXPECTED VALUES

A bettor placing a chip on a number in roulette has a small chance of winning a lot and a large chance of losing a little. To determine whether the bet is "fair," we must be able to calculate the "expected value" of the game to the bettor. An insurance company in any given year pays out a large sum of money to each of a relatively small number of people, while collecting a small sum of money from each of many other individuals. To determine policy premiums an actuary must calculate the "expected values" relating to deaths in various age groups of policy holders. In this chapter we introduce the concept of a *random variable*, and then show how to determine expected values or means and variances relating to this concept. In particular, we consider how to calculate means and standard deviations in the case of binomial probabilities.

## RANDOM VARIABLES

In Chapter 6 we determined the probability for each outcome in various examples. Often each outcome has not only an associated probability, but also an associated *real number*. For example, the probability may be 1/2 that there are 5 defective batteries; the probability may be .01 that a company receives 7 contracts; the probability may be .95 that 3 people recover from a disease. If $X$ represents the different numbers associated with the potential outcomes of some chance situation, we call $X$ a *random variable*.

**Example 7.1**  A prison official knows that 1/2 of the inmates he admits stay only 1 day, 1/4 stay 2 days, 1/5 stay 3 days, and 1/20 stay 4 days before they are either released or are sent on to the county jail. If $X$ represents the number of days, then X is a *random variable* that takes the value 1 with probability 1/2, the value 2 with probability 1/4, the value 3 with probability 1/5, and the value 4 with probability 1/20.

The random variable in Example 7.1 is called *discrete* because it can assume only a countable number of values. The random variable in Example 7.2, however, is said to be *continuous* because it can assume values associated with a whole line interval.

**Example 7.2**    Let $X$ be a random variable whose values correspond to the speeds at which a jet plane can fly. The jet may be traveling at 623.478 ... miles per hour or at any other speed in some whole interval. We might ask what the probability is that the plane is flying at between 300 and 400 miles per hour.

A *probability distribution* for a discrete variable is a listing or formula giving the probability for each value of the random variable.

**Example 7.3**    Concessionaires know that attendance at a football stadium will be 60,000 on a clear day, 45,000 if there is light snow, and 15,000 if there is heavy snow. Furthermore, the probability of clear skies, light snow, or heavy snow on any particular day is 1/2, 1/3, and 1/6 respectively. (Here we have a random variable $X$ that takes the values 60,000, 45,000, and 15,000.)

| Outcome: | clear skies | light snow | heavy snow |
|---|---|---|---|
| Probability: | 1/2 | 1/3 | 1/6 |
| Random variable: | 60,000 | 45,000 | 15,000 |

What *average* attendance should be expected for the season?

**Answer:** We reason as follows. Suppose on, say, 12 game days we had clear skies 1/2 of the time (6 days), a light snow 1/3 of the time (4 days), and a heavy snow 1/6 of the time (2 days). Then the average attendance would be

$$\frac{(6 \times 60{,}000) + (4 \times 45{,}000) + (2 \times 15{,}000)}{12} = 47{,}500$$

Note that we could have divided the 12 into each of the three terms in the numerator to obtain

$$\left(\frac{6}{12} \times 60{,}000\right) + \left(\frac{4}{12} \times 45{,}000\right) + \left(\frac{2}{12} \times 15{,}000\right)$$

or, equivalently,

$$\left(\frac{1}{2} \times 60{,}000\right) + \left(\frac{1}{3} \times 45{,}000\right) + \left(\frac{1}{6} \times 15{,}000\right) = 47{,}500$$

Actually, there was no need to consider 12 days. We could have simply multiplied probabilities times corresponding attendances and summed the resulting products.

# EXPECTED VALUE (MEAN) OF A RANDOM VARIABLE

The final calculation in Example 7.3 motivates our definition[1] of *expected value*. The *expected value* (or *average* or *mean*) of a random variable $X$ is the sum of the products obtained by multiplying each value $x$ by the corresponding probability $P(x)$:

$$E(X) = \Sigma\, x\, P(x)$$

**Example 7.4**     In a lottery, 10,000 tickets are sold at $1 each with a prize of $7500 for one winner. What is the average result for each bettor?

*Answer:* The actual winning payoff is $7499 because the winner paid $1 for a ticket, so we have:

| Outcome: | win | lose |
|---|---|---|
| Probability: | $\dfrac{1}{10,000}$ | $\dfrac{9,999}{10,000}$ |
| Random variable: | 7499 | $-1$ |

$$\text{Expected value} = 7499\left(\frac{1}{10,000}\right) + (-1)\left(\frac{9,999}{10,000}\right) = -0.25.$$

Thus the *average* result for each person betting the lottery is a 25¢ loss.

**Example 7.5**     A manager must choose among three options. Option $A$ has a 10% chance of resulting in a $250,000 gain, but otherwise will result in a $10,000 loss. Option $B$ has a 50% chance of gaining $40,000 and a 50% chance of losing $1000. Finally, option $C$ has a 5% chance of gaining $800,000, but otherwise will result in a loss of $20,000. Which option should the manager choose?

---

[1]The development here is limited to random variables with a finite number of values, but this definition can be extended somewhat using infinite sums.

*Answer:* Calculate the expected values of the three options:

|  | Option A | | Option B | | Option C | |
|---|---|---|---|---|---|---|
| Outcome: | gain | loss | gain | loss | gain | loss |
| Probability: | .10 | .90 | .50 | .50 | .05 | .95 |
| Random variable: | 250,000 | −10,000 | 40,000 | −2000 | 800,000 | −20,000 |

$$E(A) = .10(250,000) + .90(-10,000) = \$16,000$$
$$E(B) = .50(40,000) + .50(-2000) = \$19,000$$
$$E(C) = .05(800,000) + .95(-20,000) = \$21,000$$

The manager should choose option *C*! However, although Option *C* has the greatest mean (expected value), the manager may well wish to consider the relative riskiness of the various options. If, for example, a $5000 loss would be disastrous for the company, the manager might well decide to choose option *B* with its maximum possible loss of $2000. (It should be intuitively clear that some concept of variance would be helpful here in measuring the risk; later in this chapter the variance of a random variable will be defined.)

Suppose we have a *binomial random variable*, that is, a random variable whose values are the numbers of "successes" in some binomial probability distribution.

**Example 7.6**    Of the automobiles produced in a particular plant, 40% had a certain defect. Suppose a company purchases five of these cars. What is the expected value for the number of cars with defects?

*Answer:* We might guess that the average or mean or expected value is 40% of 5 = .4 × 5 = 2, but let's calculate from the definition. Letting *X* represent the number of cars with the defect, we have:

$$P(0) = C(5,0)(.4)^0(.6)^5 = \quad (.6)^5 \quad = .07776$$
$$P(1) = C(5,1)(.4)^1(.6)^4 = 5(.4)(.6)^4 = .25920$$
$$P(2) = C(5,2)(.4)^2(.6)^3 = 10(.4)^2(.6)^3 = .34560$$
$$P(3) = C(5,3)(.4)^3(.6)^2 = 10(.4)^3(.6)^2 = .23040$$
$$P(4) = C(5,4)(.4)^4(.6)^1 = 5(.4)^4(.6) = .07680$$
$$P(5) = C(5,5)(.4)^5(.6)^0 = \quad (.4)^5 \quad = .01024$$

| Outcome: | 0 cars | 1 car | 2 cars | 3 cars | 4 cars | 5 cars |
|---|---|---|---|---|---|---|
| Probability: | .07776 | .25920 | .34560 | .23040 | .07680 | .01024 |
| Random variable: | 0 | 1 | 2 | 3 | 4 | 5 |

$$E(X) = 0(.07776) + 1(.25920) + 2(.34560) + 3(.23040) + 4(.07680) + 5(.01024) = 2$$

Thus, the answer turns out to be the same as would be obtained by simply multiplying the probability of "success" times the number of cases.

The following is true: If we have a binomial probability situation with the probability of success equal to $p$ and the number of trials equal to $n$, the *expected value* or *mean* number of successes for the $n$ trials is $np$.

One proof of this statement involves algebraic manipulation with the binomial probability formula to show that $\Sigma xP(x) = np$. Another insight is as follows:

If $n = 1$ we have:

| Outcome: | success | failure |
|---|---|---|
| Probability: | $p$ | $1-p$ |
| Random Variable (successes): | 1 | 0 |

with expected value $= 1(p) + 0(1 - p) = p$. For larger $n$, we consider each trial independently with a resulting expected value of $p$ for each trial. The number of successes in $n$ trials is the sum of the successes in all $n$ trials, and so the expected value or mean of the $n$ trials is the sum of the expected values of all $n$ trials. We calculate

$$p + p + p + \ldots + p = np$$

One $p$ for each of $n$ trials

**Example 7.7**   An insurance salesperson is able to sell policies to 15% of the people she contacts. Suppose that she contacts 120 people during a 2-week period. What is the expected value for the number of policies she sells?

***Answer:*** We have a binomial probability with the probability of success .15 and the number of trials 120, so the mean or expected value for the number of successes is $120 \times .15 = 18$.

# VARIANCE OF A RANDOM VARIABLE

We have seen that the *mean* of a random variable is $\Sigma xP(x)$. However, not only is the mean important, but also we would like to measure the *variability* for the values taken on by a random variable. Since we are dealing with chance events, the proper tool is *variance*. Variance was defined in Chapter 2 to be the mean average of the squared deviations $(x - \mu)^2$. If we regard the $(x - \mu)^2$ terms as the values of some random variable (whose probability is the same as the probability of $x$), then the mean of this new random variable is

$\Sigma (x-\mu)^2 P(x)$, which is precisely how we define the variance $\sigma^2$ of a random variable:

$$\sigma^2 = \Sigma(x - m)^2 P(x)$$

As before, the standard deviation $\sigma$ is the square root of the variance.

**Example 7.8**    A highway engineer knows that his crew can lay 5 miles of highway on a clear day, 2 miles on a rainy day, and only 1 mile on a snowy day. Suppose the probabilities are as follows:

| Outcome: | clear | rain | snow |
|---|---|---|---|
| Probability: | .6 | .3 | .1 |
| Random Variable (miles of highway): | 5 | 2 | 1 |

What are the mean (expected value) and the variance?

***Answer:***

$$\mu = \Sigma\, xP(x) = 5(.6) + 2(.3) + 1(.1) = 3.7$$

$$\sigma^2 = \Sigma(x - \mu)^2 P(x)$$
$$= (5 - 3.7)^2(.6) + (2 - 3.7)^2(.3) + (1 - 3.7)^2(.1) = 2.61$$

**Example 7.9**    Look again at Example 7.6. We calculated the mean to be 2. What is the variance?

***Answer:*** $\sigma^2 = (0-2)^2(.07776) + (1-2)^2(.2592) + (2-2)^2(.3456)$
$+ (3-2)^2(.2304) + (4-2)^2(.0768) + (5-2)^2(.01024) = 1.2$

Could we have calculated the above result more easily? In this case, we have a binomial random variable, so we will use the same type of argument that we used in showing that $\mu = np$. We first consider the binomial distribution for $n = 1$:

| Outcome: | success | failure |
|---|---|---|
| Probability: | $p$ | $q$ |
| Random variable (number of successes): | 1 | 0 |

Now $\mu = p(1) + q(0) = p$, so

$$\begin{aligned}
\sigma^2 &= (1 - \mu)^2 p + (0 - \mu)^2 q & \\
&= q^2 p + p^2 q & [\mu = p \text{ and } 1 - p = q] \\
&= pq(p + q) & [\text{factoring}] \\
&= pq & [p + q = 1]
\end{aligned}$$

For larger $n$, we reason as follows. The number of successes is the sum of the successes in all trials, so the variance is the sum of the variances from all trials (Chapter 2):

$$pq + pq + pq + \dots + pq = npq$$

one $pq$ for each of $n$ trials

More formally, this result can be obtained through algebraic manipulations (see Study Question 3).

**Example 7.10**  How can we use this method to more simply calculate the variance in Example 7.9.

**Answer:** $npq = 5(.4)(.6) = 1.2.$

Thus, for a random variable $X$,

Mean or expected value    $\mu = \Sigma\, x\, P(x)$

Variance    $\sigma^2 = \Sigma(x - \mu)^2\, P(x)$

Standard deviation    $\sigma = \sqrt{\Sigma(x - \mu)^2\, P(x)}$

In the case of a binomial probability distribution with probability of success equal to $p$, and number of trials equal to $n$, if we let $X$ be the number of successes in the $n$ trials, the above equations become:

Mean or expected value    $\mu = np$

Variance    $\sigma^2 = npq = np(1 - p)$

Standard deviation    $\sigma = \sqrt{npq} = \sqrt{np(1 - p)}$

**Example 7.11**  Sixty percent of all new car buyers choose automatic transmissions. For a group of five new car buyers, calculate the mean and standard deviation for the number of buyers choosing automatics.

**Answer:**

$$\mu = np = 5(.6) = 3.0$$

$$\sigma = \sqrt{np(1 - p)} = \sqrt{5(.6)(.4)} = 1.1$$

Note that this could have been calculated through the more involved

$$\mu = \Sigma\, x\, P(x) = 0[(.4)^5] + 1[5(.6)(.4)^4] + 2[10(.6)^2(.4)^3]$$
$$+ 3[10(.6)^3(.4)^2] + 4[5(.6)^4(.4)] + 5[(.6)^5] = 3.0$$

$$\sigma = \sqrt{\Sigma(x - \mu)^2 P(x)}$$
$$= \sqrt{9[.01024] + 4[.07680] + 1[.23040] + 0[.34560] + 1[.25920] + 4[.07776]}$$
$$= 1.1$$

# EXERCISES

1. Alan Dershowitz, one of O.J. Simpson's lawyers, has stated that only one out of every 1000 abusive relationships ends in murder each year. If he is correct, and if there are approximately 1,500,000 abusive relationships in the United States, what is the expected value for the number of people who are killed each year by their abusive partners?

2. If 3% of the population is allergic to the malaria fighting drug chloroquine, what is the expected value for the number of allergic people in a town of 1200 people?

3. A television game show has three payoffs with the following probabilities:

| Payoff ($): | 0 | 1000 | 10,000 |
|---|---|---|---|
| Probability: | .6 | .3 | .1 |

   What are the mean and standard deviation for the payoff variable?

4. Companies proved to have violated pollution laws are being fined various amounts with the following probabilities:

| Fine ($) | 1000 | 10,000 | 50,000 | 100,000 |
|---|---|---|---|---|
| Probability | .4 | .3 | .2 | .1 |

   What are the mean and standard deviation for the fine variable?

5. *The New York Times* (September 21, 1994) reported that an American woman diagnosed with ovarian cancer has a 37.5% chance of survival, and that approximately 20,000 American women are diagnosed with ovarian cancer each year.
   a. What is the expected number of deaths annually from this disease?
   b. What are the mean, variance, and standard deviation for a binomial with $n = 20,000$ and $p = .625$?

6. Of the coral reef species in the Great Barrier Reef off Australia, 73% are poisonous. If a tourist boat bringing divers to different points off the reef encounters an average of 25 coral reef species, what are the mean and standard deviation for the expected number of poisonous species seen?

## Study Questions

1. A famous problem asks you to choose between two envelopes, one of which has twice as much money as the other. You arbitrarily pick one, open it, and find $100. You are then given the chance to switch envelopes. You reason that the other envelope has either $50 or $200, each with a .5 probability. Applying your understanding of expected value, you calculate .5($50)+.5($200) = $125 and conclude that you should switch envelopes. Comment on this reasoning.

2. Show algebraically that in the case of a binomial distribution, the formula $\mu = \Sigma x\, P(x)$ reduces to $np$. *Hint*:

$$0[(1-p)^n] + 1[np(1-p)^{n-1}] + 2\left[\frac{n(n-1)}{2}\, p^2(1-p)^{n-2}\right]$$
$$+ 3\left[\frac{n(n-1)(n-2)}{3!}\, p^3(1-p)^{n-3}\right] + \ldots + n[p^n]$$
$$= np[(1-p)^{n-1} + (n-1)(1-p)^{n-2}p + \frac{(n-1)(n-2)}{2!}(1-p)^{n-3}p^2$$
$$+ \ldots + p^{n-1}] = np[(1-p) + p]^{n-1} = \ldots$$

3. Show algebraically that, in the case of a binomial distribution with $n = 2$, the formula $\sigma^2 = \Sigma(x-\mu)^2\, P(x)$ reduces to $np(1-p)$. [*Hint*: Use the result from Study Question 2 that $\mu = np$.] How about a binomial distribution with $n = 3$?

# Part 3

# PROBABILITY DISTRIBUTIONS

*With an understanding of binomial probabilities and expected values, we can now develop and explore several probability distributions of general interest in statistics. Such knowledge not only is necessary for our future study, but also is immediately useful as a decision making tool for certain classes of problems.*

*For example, knowing the probability of finding oil given certain geological conditions, we can use the* binomial distribution *to calculate the probabilities of various numbers of positive strikes for a given number of test sites. Knowing the average number of Supreme Court vacancies during previous presidential terms, we can use the* Poisson distribution *to calculate the probabilities of various numbers of vacancies arising during the next 4-year term. Knowing the mean and variance of heights of U.S. Marines, we can use the* normal distribution *to calculate the probability that any Marine has a height greater than a specified value.*

*Our discussion in Chapter 8 of the properties of the binomial distribution arises from the concepts and calculations of Chapters 6 and 7. The Poisson distribution in Chapter 9 can be viewed as a limiting case of the binomial when* n *is large and* p *is small, and the normal distribution in Chapter 10 as a limiting case of the binomial when* p *is constant but* n *increases without bound. With this approach, it will be clear that both the Poisson and the normal distributions can be used as approximations to binomial problems.*

*Finally, it cannot be overstressed that, even if there were very few naturally occurring normal distributions, the normal has tremendous value in that it describes the distribution found in a wide range of statistical experiments, investigations, and studies. This application will be the focus of Parts 4 and 5.*

# 8
# BINOMIAL DISTRIBUTIONS

Suppose an experiment or situation has a finite number of possible outcomes. Each outcome has an associated probability, and there are various ways to display these probabilities. Two primary ways are tables and histograms.

**Example 8.1**  A switchboard operator notes that for any 1-minute period, the probabilities for no, one, two, and three calls are .3, .4, .2, and .1, respectively. We display these probabilities as follows:

**IN A TABLE**  and  **IN A HISTOGRAM**

| Number of Calls | Probability |
|:---:|:---:|
| 0 | .3 |
| 1 | .4 |
| 2 | .2 |
| 3 | .1 |

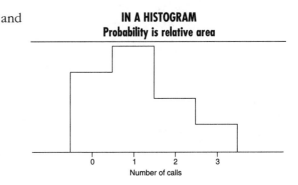

Probability is relative area

Number of calls

The set of probabilities {.3, .4, .2, .1} in Example 8.1 is called a *probability distribution*. Over a large number of 1-minute periods, we would expect no calls in approximately 30% of the periods, one call in 40%, two calls in 20%, and three calls in 10%. The larger the number of 1-minute periods considered, the closer we would expect the observed percentage to be to these theoretical percentages.

In this chapter we consider probability distributions arising from binomial probabilities. The concept of binomial probability developed in Chapter 6 can be expanded in various directions. We will consider tables, histograms, and the notions of mean and variance, and will then look at several applications.

To review, a binomial experiment has the following characteristics:

1. The experiment consists of $n$ identical trials.
2. Each trial has the same two outcomes, commonly called *success* and *failure*, with probabilities $p$ and $1 - p$, respectively.
3. The probabilities of success and failure remain the same from trial to trial; that is, the outcome of any one trial has no effect on the outcome of any other trial.
4. The mean is $\mu = np$.
5. The standard deviation is $\sigma = \sqrt{np(1 - p)}$

**Example 8.2**   The probability is 1/3 that a supermarket customer will spend over $50. Suppose that there are four customers at a checkout counter. What is the probability distribution for the number of these customers spending over $50?

**Answer:**

$$
\begin{aligned}
P(0) &= C(4,0)(1/3)^0(2/3)^4 = & 1(2/3)^4 & = .198 \\
P(1) &= C(4,1)(1/3)^1(2/3)^3 = & 4(1/3)(2/3)^3 & = .395 \\
P(2) &= C(4,2)(1/3)^2(2/3)^2 = & 6(1/3)^2(2/3)^2 & = .296 \\
P(3) &= C(4,3)(1/3)^3(2/3)^1 = & 4(1/3)^3(2/3) & = .099 \\
P(4) &= C(4,4)(1/3)^4(2/3)^0 = & 1(1/3)^4 & = .012 \\
& & & \overline{1.000}
\end{aligned}
$$

| Customers spending over $50 | Probability |
|---|---|
| 0 | .198 |
| 1 | .395 |
| 2 | .296 |
| 3 | .099 |
| 4 | .012 |

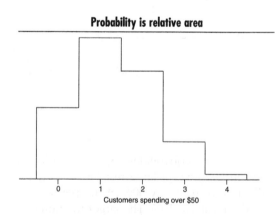

**Probability is relative area**

Customers spending over $50

The mean and standard deviation of the probability distribution in Example 8.2 are as follows:

$$
\mu = np = (4)\left(\frac{1}{3}\right) = 1.333 \qquad \sigma = \sqrt{np(1 - p)} = \sqrt{4\left(\frac{1}{3}\right)\left(\frac{2}{3}\right)} = .943
$$

**Example 8.3**  The probability is .6 that a well driller will find water at a depth of less than 100 feet in a certain area. Wells are to be drilled for six new home owners. What is the complete probability distribution for the number of wells under 100 feet?

*Answer:* Using the appropriate line (1-6-15-20-15-6-1) from Pascal's triangle (Chapter 5), we have:

$$P(0) = 1(.6)^0(.4)^6 = .004$$
$$P(1) = 6(.6)^1(.4)^5 = .037$$
$$P(2) = 15(.6)^2(.4)^4 = .138$$
$$P(3) = 20(.6)^3(.4)^3 = .276$$
$$P(4) = 15(.6)^4(.4)^2 = .311$$
$$P(5) = 6(.6)^5(.4)^1 = .187$$
$$P(6) = 1(.6)^6(.4)^0 = \underline{.047}$$
$$1.000$$

| Wells under 100 feet | Probability |
|:---:|:---:|
| 0 | .004 |
| 1 | .037 |
| 2 | .138 |
| 3 | .276 |
| 4 | .311 |
| 5 | .187 |
| 6 | .047 |

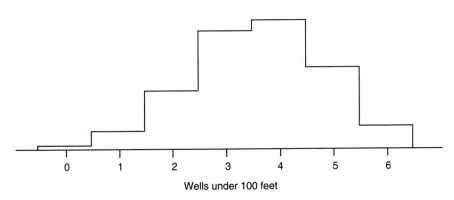

Wells under 100 feet

The mean and standard deviation of this probability distribution are calculated as follows:

$$\mu = np = 6(.6) = 3.6 \qquad \sigma = \sqrt{np(1-p)} = \sqrt{6(.6)(.4)} = 1.2$$

It is useful to also label the horizontal axis with $z$-scores (see Chapter 4). In this case, $z$-scores of 0, 1, and 2 correspond to 3.6, 4.8, and 6.0, while $z$-scores of $-1$, $-2$, and $-3$ correspond to 2.4, 1.2, and 0, respectively.

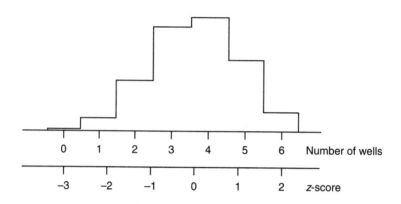

**Example 8.4**   A baseball player with a batting average of .250 has 12 official at-bats in a three-game series. What is the probability distribution for the number of hits he makes? Display in a histogram.

***Answer:*** The probability of a hit is .250, and we have:

$$
\begin{aligned}
P(0) &= C(12,0)(.250)^0(.750)^{12} = & 1(.75)^{12} &= .032 \\
P(1) &= C(12,1)(.250)^1(.750)^{11} = & 12(.25)(.75)^{11} &= .127 \\
P(2) &= C(12,2)(.250)^2(.750)^{10} = & 66(.25)^2(.75)^{10} &= .232 \\
P(3) &= C(12,3)(.250)^3(.750)^9 = & 220(.25)^3(.75)^9 &= .258 \\
P(4) &= C(12,4)(.250)^4(.750)^8 = & 495(.25)^4(.75)^8 &= .194 \\
P(5) &= C(12,5)(.250)^5(.750)^7 = & 792(.25)^5(.75)^7 &= .103 \\
P(6) &= C(12,6)(.250)^6(.750)^6 = & 924(.25)^6(.75)^6 &= .040 \\
P(7) &= C(12,7)(.250)^7(.750)^5 = & 792(.25)^7(.75)^5 &= .012 \\
P(8) &= C(12,8)(.250)^8(.750)^4 = & 495(.25)^8(.75)^4 &= .002
\end{aligned}
$$

To three decimal places $P(9) = P(10) = P(11) = P(12) = 0$. The mean number of hits is

$$\mu = np = 12(.25) = 3$$

and the standard deviation is

$$\sigma = \sqrt{np(1 - p)} = \sqrt{12(.25)(.75)} = 1.5$$

Thus, $z$-scores of 0, 1, and 2 correspond to 3, 4.5, and 6, while $z$-scores of $-1$ and $-2$ correspond to 1.5 and 0, respectively.

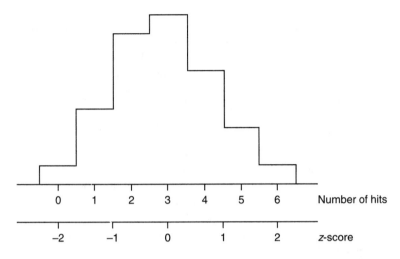

Note how the value of $p$ affects the shape of the distribution as displayed in a histogram. For $n = 10$, the following are the histograms for $p = .5$, $p = .1$, and $p = .7$.

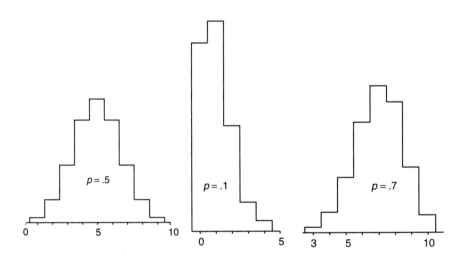

For $p = .5$ the distribution is symmetric with the greatest probability occurring at the center. For $p = .7$ the high point of the histogram is shifted to the right, and for $p = .1$ it is shifted even farther to the left.

The following examples and exercises illustrate direct applications of the binomial distribution.

**Example 8.5**    A common decision-making problem arising in manufacturing situations involves whether to accept a shipment of raw materials

or of finished products. Concerns over quality control are sometimes handled by inspecting a sample before accepting a whole
shipment. The shipment is accepted only if fewer than some
specified number of defects are found in the sample. Decisions on
sample size and allowable number of defects are specific to the
given situation and must take into consideration time, money,
quality needs, and other relevant factors.

In analyzing a proposed sampling plan, it is important to calculate
the probability of shipment acceptance given various possible
levels of product defect. For example, suppose the decision-
making rule is to pick a sample of size $n = 10$, and accept the
whole shipment if the sample contains at most one defective item.
What is the probability of acceptance if the defect level of the
shipment is actually 5%? 10%? 20%? 30%? 40%? 50%?

**Answer:**

For $p = .05$,

$$P \text{ (accept)} = P(0 \text{ def}) + P(1 \text{ def})$$

$$= C(10,0)(.05)^0(.95)^{10} + C(10,1)(.05)^1(.95)^9$$

Thus we have:

For $p = .05$, $P(\text{accept}) = (.95)^{10} + 10(.05)(.95)^9 = .914$
For $p = .10$, $P(\text{accept}) = (.90)^{10} + 10(.10)(.90)^9 = .736$
For $p = .20$, $P(\text{accept}) = (.80)^{10} + 10(.20)(.80)^9 = .376$
For $p = .30$, $P(\text{accept}) = (.70)^{10} + 10(.30)(.70)^9 = .149$
For $p = .40$, $P(\text{accept}) = (.60)^{10} + 10(.40)(.60)^9 = .046$
For $p = .50$, $P(\text{accept}) = (.50)^{10} + 10(.50)(.50)^9 = .011$

The graph of the above values is called the *operating characteristic curve* for this sampling plan.

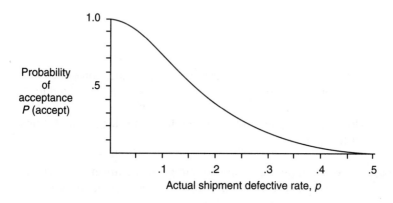

Probability
of
acceptance
P (accept)

Actual shipment defective rate, p

Note that, if $p = 0$, that is, if there is no defective item in the entire shipment, then $P(\text{accept}) = 1$. On the other hand, if $p = 1.0$, that is, if the entire shipment is defective, then $P(\text{accept}) = 0$.

**Example 8.6**  We can test the randomness of certain happenings by observing how closely the results follow an expected binomial pattern. For example, suppose there are 100 accounting firms each with 10 new employees. We want to test the hypothesis that the new employees joined the firms randomly, and for data we have results from the CPA exam taken by all the new employees. Suppose that in 2 firms no new employee passed the exam, in 6 firms one new employee passed, in 13 firms two new employees passed, in 20 three passed, in 28 four passed, in 16 five passed, in 11 six passed, in 4 seven passed, and no firms have more than seven new employees who passed the CPA exam.

| Number of new employees who passed: | 0 | 1 | 2 | 3 | 4 | 5 | 6 | 7 | 8 | 9 | 10 |
|---|---|---|---|---|---|---|---|---|---|---|---|
| Number of firms: | 2 | 6 | 13 | 20 | 28 | 16 | 11 | 4 | 0 | 0 | 0 |

Would the above results be likely if the new employees joined the various firms *randomly*.

**Answer:** If these new employees were distributed at random among the firms, what would be the distribution with regard to number of new employees who passed the CPA exam? First, we determine the probability, $p$, of passing the exam by summing the number of the 1000 new employees who passed and then dividing by 1000.

$$p = \frac{(2 \times 0) + (6 \times 1) + (13 \times 2) + (20 \times 3) + (28 \times 4) + (16 \times 5) + (11 \times 6) + (4 \times 7)}{1000}$$

$$= .378$$

Now, what is the probability that a firm had none of its ten new employees pass? One new employee pass? Two out of ten pass? Three? Four? We recognize this question as a binomial with $p = .378$ and $n = 10$. Using the appropriate line from Pascal's triangle, 1-10-45-120-210-252-210-120-45-10-1, we calculate:

$$
\begin{aligned}
P(0) &= (.622)^{10} &= .009 \\
P(1) &= 10(.378)(.622)^9 &= .053 \\
P(2) &= 45(.378)^2(.622)^8 &= .145 \\
P(3) &= 120(.378)^3(.622)^7 &= .233 \\
P(4) &= 210(.378)^4(.622)^6 &= .248 \\
P(5) &= 252(.378)^5(.622)^5 &= .181 \\
P(6) &= 210(.378)^6(.622)^4 &= .092
\end{aligned}
$$

$$
\begin{aligned}
P(7) &= 120(.378)^7(.622)^3 &&= .032 \\
P(8) &= 45(.378)^8(.622)^2 &&= .007 \\
P(9) &= 10(.378)^9(.622) &&= .001 \\
P(10) &= (.378)^{10} &&= .000
\end{aligned}
$$

Multiplying each of these probabilities by 100 (there are 100 firms) gives the "expected" number of firms with 0 employees passing, one employee passing, two employees passing, and so on.

| Number of new employees who passed: | 0 | 1 | 2 | 3 | 4 | 5 | 6 | 7 | 8 | 9 | 10 |
|---|---|---|---|---|---|---|---|---|---|---|---|
| Expected number of firms: | | 1 | 5 | 15 | 23 | 25 | 18 | 9 | 3 | 1 | 0 | 0 |
| Actual number of firms: | | 2 | 6 | 13 | 20 | 28 | 16 | 11 | 4 | 0 | 0 | 0 |

Are the "expected" and "actual" numbers close enough to justify the conclusion that we have a binomial distribution, and thus that the new employees who passed the CPA exam are randomly distributed among firms? The numbers do appear to be very close, and we probably would conclude that the variations in exam results between firms were due to random influences.

In Chapter 19 on chi-square analysis for goodness-of-fit tests, we will discuss a mathematical technique for measuring the strength of agreement between two such sets of numbers. Meanwhile, the type of analysis illustrated in Example 8.6 is standard and useful. A similar argument in the next chapter will use the *Poisson distribution* instead of the binomial.

# EXERCISES

1. It has been shown that in major league baseball games, home teams win approximately 55% of the time. Suppose a team plays a three game home series. What is the probability distribution for the number of wins?

2. Suppose USAir accounts for 20% of U.S. domestic flights.
   a. If there are seven major air disasters in the United States, what is the probability distribution for the number of these disasters for which USAir might be accountable?
   b. What is the probability that USAir accounts for at least four of the disasters? (USAir actually did account for four of the last seven major disasters. "That's enough to begin getting suspicious but not enough to hang them," said Dr. Brad Efton in *The New York Times*, September 11, 1994.)

3. A study reported by Liu and Schutz states that in hockey the better team has a 65% chance of scoring the first goal in overtime.

   a. If a team is considered to be the better team in all five of its overtime games, find the probability distribution for the number of these games in which that team will score the first overtime goal.
   b. Find the mean and standard deviation for this distribution.
   c. Draw a histogram showing two horizontal axes, one for number of first overtime goals, and one for $z$-scores.

4. *The New York Times*, June 21, 1993, reported that one out of every three people in the world is infected with tuberculosis (although most cases are latent).
   a. For a group of ten people, find the probability distribution for the number who may have tuberculosis.
   b. Find the mean and standard deviation for this distribution.
   c. Draw a histogram showing two horizontal axes, one for number of people with tuberculosis, and one for $z$-scores.

5. An inspection procedure at a manufacturing plant involves picking three items at random and then accepting the whole lot if at least two of the three items are in perfect condition. If in reality 90% of the whole lot are perfect, what is the probability that the lot will be accepted? If 75% of the whole lot are perfect? If 60% are perfect?

## Study Questions

1. Seeds of a new variety are planted in 50 rows with eight seeds to a row. The number of germinations per row is noted, with the following results:

| Number of germinations: | 0 | 1 | 2 | 3 | 4 | 5 | 6 | 7 | 8 |
|---|---|---|---|---|---|---|---|---|---|
| Actual number of rows: | 0 | 0 | 1 | 2 | 5 | 8 | 18 | 12 | 5 |

   a. How many of the 400 seeds germinated?
   b. What is the probability that a seed germinates?
   c. What is the binomial probability distribution with this probability and with $n = 8$?
   d. Multiplying each probability in this distribution by 50 gives how many expected numbers of rows for each number of germinations?
   e. Do the expected numbers seem to be close enough to the actual numbers that we can conclude the seeds were randomly distributed among the rows?

2. A new antibiotic is tested on six patients in each of 125 hospitals. The number of positive responses among the patients in each hospital is noted with the following results:

| Number of positive responses: | 0 | 1 | 2 | 3 | 4 | 5 | 6 |
|---|---|---|---|---|---|---|---|
| Actual number of hospitals: | 0 | 0 | 5 | 36 | 51 | 20 | 13 |

   a. How many patients had a positive response?
   b. What is the probability that a patient will have a positive response?

c. What is the binomial probability distribution with this probability and with $n = 6$?

d. What is the expected number of hospitals for each number of positive responses?

e. Do the expected and actual numbers of hospitals seem to be close? (If the actual numbers are close to the numbers predicted by the binomial distribution, we will be more confident in concluding that the results are due solely to the antibiotic and not to differences in the hospitals and other factors.)

# 9
# POISSON DISTRIBUTIONS

To calculate a binomial distribution, we must know $n$, the number of cases under consideration. In other words, to find the probability of a certain number of successes, we must also know the corresponding number of failures. However, there are situations in which the number of failures cannot be determined. For example, in calculating the probability that a hospital will have two appendectomy cases in 1 day, it makes little sense to ask how many appendicitis cases will not occur. In calculating the probability that a baseball team will score five runs in a game, it is meaningless to ask how many runs will not be scored. And in finding the probability that there will be ten incoming telephone calls at a switchboard, it is impossible to say how many calls will not come in.

In a wide variety of examples, both $n$, the number of cases, and $p$, the probability of success, are unknown. In this chapter we consider the special cases in which $n$ is very large, and $p$ is very small, but the mean or average $\mu = np$ is both moderate and known. For example, a large number $n$ of planes fly into a major airport, and the probability $p$ that any particular plane will be on the runway at any specific moment in time is small, however the average number $np$ of planes waiting in line on the runway at noon each day can be determined without evaluating either $p$ or $n$. Similarly, a large number $n$ of machine parts could pass through a distribution center each day, and the probability $p$ that any particular item would be defective might be small; however, the average number $np$ of defective items found each day could be noted without knowing either $p$ or $n$. For such situations we develop a technique for calculating the probabilities of various numbers of "successes."

## PROBABILITY OF ZERO SUCCESSES

First we consider the problem of determining the probability of no successes. We have seen that for a binomial experiment with $n$ cases and probability of success $p$, the probability of one failure is $1 - p$, and the probability of $n$ consecutive failures is $(1 - p)^n$. We introduce the mean $\mu = np$ algebraically by noting that

$$n = \frac{1}{p}(pn) = \frac{1}{p}\mu \qquad so\ (1 - p)^n = (1 - p)^{(1/p)\mu} = \left[(1 - p)^{1/p}\right]^{\mu}$$

What can be said about $(1 - p)^{1/p}$ in our special case concerned with a small value of $p$?

A hand calculator gives the following results:

| $p$: | .5 | .1 | .001 | .00001 | .0000000001 |
|---|---|---|---|---|---|
| $(1-p)^{1/p}$: | .25000 | .34868 | .36770 | .36788 | .36788 |

Mathematically, it can be shown that, as $p$ is chosen to be closer and closer to 0, $(1-p)^{1/p}$ is closer and closer to $1/e$, where $e$ is a mathematical constant that arises in a wide variety of mathematical fields and applications.

$$e = 2.71828\ldots \qquad \text{and} \qquad \frac{1}{e} = .36788\ldots$$

Thus, for small $p$, the probability of no successes,

$$\left[(1-p)^{1/p}\right]^{\mu}$$

is approximately $(1/e)^{\mu}$ or $e^{-\mu}$.

**Example 9.1**    A supermarket manager knows that an average of four persons will pass through the checkout line during any 10 minute period.
a. What is the probability of no customer in a 10 minute period?

***Answer:*** $e^{-4} = .018$

b. What is the probability of no customer in a five-minute period? In a 1-minute period?

***Answer:*** If the average in 10 minutes is four persons, then the average in 5 minutes is $4/2 = 2$ persons, and the average in 1 minute is $4/10 = 0.4$ people. The probability of no customer in a 5-minute period is thus $e^{-2} = .135$. In a 1-minute period the probability is $e^{-0.4} = .670$.

**Example 9.2**    A national highway study claims that an average of 1.7 accidents occur for every 100-mile length of interstate highway every year.
a. What is the probability of no accident on any 100-mile stretch?

***Answer:*** $e^{-1.7} = .183$

b. What is the probability of no accident in a 200-mile stretch of interstate highway? In a 10-mile stretch?

***Answer:*** In a 200-mile stretch the yearly average will be $2(1.7) = 3.4$ accidents, so the probability of no accident is $e^{-3.4} = .033$. For a 10-mile stretch the average is $1.7/10 = 0.17$ accident, and so the probability of no accident is $e^{-0.17} = .844$.

# PROBABILITIES OF POSITIVE NUMBERS OF SUCCESSES

The key to proceeding from the probability of no success to the probabilities of one or more successes is the binomial formula from Chapter 6:

$$P(x \; successes) = \frac{n - x + 1}{x} \frac{p}{1 - p} P(x - 1 \; successes)$$

In our case $p$ is small so $1 - p$ is approximately equal to 1, and $n$ is large so $n-x+1$ is approximately equal to $n$. In this situation

$$\frac{n - x + 1}{x} \frac{p}{1 - p} \approx \frac{n}{x} \frac{p}{1} = \frac{np}{x} = \frac{\mu}{x}$$

and our formula reduces to

$$P(x \; successes) = \frac{\mu}{x} P(x - 1 \; successes)$$

Combining this formula with the derivation for the probability of no success leads to the following:

$$P(0 \; successes) = e^{-\mu}$$

$$P(1 \; success) = \frac{\mu}{1} e^{-\mu}$$

$$P(2 \; successes) = \frac{\mu}{2} \left[ \frac{\mu}{1} e^{-\mu} \right] = \frac{\mu^2}{2} e^{-\mu}$$

$$P(3 \; successes) = \frac{\mu}{3} \left[ \frac{\mu^2}{2} e^{-\mu} \right] = \frac{\mu^3}{3!} e^{-\mu}$$

$$P(4 \; successes) = \frac{\mu}{4} \left[ \frac{\mu^3}{3!} e^{-\mu} \right] = \frac{\mu^4}{4!} e^{-\mu}$$

and, more generally,

$$P(x \; successes) = \frac{\mu^x}{x!} e^{-\mu}$$

**Example 9.3** In a large northeastern town there is an average of 1.5 business bankruptcies per week. What is the probability of no bankruptcy in a week? One bankruptcy? Two bankruptcies? Three?

*Answer:* There are a large number of businesses, the probability of bankruptcy is small, and the average is known. Therefore, we apply the formula we developed to obtain:

$$P(0 \text{ bankruptcies}) = e^{-1.5} \qquad = .223$$

$$P(1 \text{ bankruptcy}) \quad = 1.5e^{-1.5} \qquad = .335$$

$$P(2 \text{ bankruptcies}) = [(1.5)^2/2]e^{-1.5} = .251$$

$$P(3 \text{ bankruptcies}) = [(1.5)^3/3\,!]e^{-1.5} = .126$$

**Example 9.4**    A grocery store manager knows that his store sells an average of three cans of artichoke hearts per week. Assuming that this situation is described by the formula developed above, what is the probability that no can of artichoke hearts will be sold in 1 week? The probability of one can in 1 week? Two cans in 1 week? One can in 1/2 week? Three cans in 2 weeks?

***Answer:***

For 1 week the average is three cans, so

$$P(0 \text{ cans in 1 week}) \quad = e^{-3} \qquad = .050$$

$$P(1 \text{ can in 1 week}) \quad = 3e^{-3} \qquad = .149$$

$$P(2 \text{ cans in 1 week}) \quad = (3^2/2)e^{-3} \quad = .224$$

For 1/2 week the average is $.5(3) = 1.5$, so

$$P(1 \text{ can in 1/2 week}) \quad = 1.5e^{-1.5} \qquad = .335$$

For 2 weeks the average is $2(3)$ $\qquad = 6$, so

$$P(3 \text{ cans in 2 weeks}) \quad = [6^3/3\,!]e^{-6} \quad = .089$$

# POISSON PROBABILITY DISTRIBUTION

A probability distribution given by

$$P(x \text{ successes}) = \frac{\mu^x}{x\,!}\, e^{-\mu}$$

where $\mu$ is the mean and $e = 2.71828...$, is called a *Poisson probability distribution*. We have seen that a Poisson distribution will result as the limiting case of a binomial distribution for large $n$, small $p$, and moderate $\mu$. The Poisson is often the appropriate distribution description in situations in which the probability of one occurrence during a short interval is proportional to the length of that interval.

**Example 9.5**    One of the first noted examples of a Poisson distribution involves the number of deaths from horse kicks in various corps of the

German army in the late 1800s. More specifically, during the 20-year period from 1875 to 1894, among the 14 cavalry corps of the German army, there was an average 0.7 deaths per corps per year as the result of horse kicks. If the deaths followed a Poisson distribution, the descriptive probabilities per corp per year would be as follows:

$$P(0 \text{ deaths}) = e^{-0.7} \qquad\qquad = .497$$
$$P(1 \text{ death}) = 0.7e^{-0.7} \qquad\quad = .348$$
$$P(2 \text{ deaths}) = [(0.7)^2/2]e^{-0.7} = .122$$
$$P(3 \text{ deaths}) = [(0.7)^3/3\,!]e^{-0.7} = .028$$
$$P(4 \text{ deaths}) = [(0.7)^4/4\,!]e^{-0.7} = .005$$
$$\overline{\phantom{xxxxx}1.000}$$

If these are the correct probabilities, then the expected numerical distribution, found by multiplying the probabilities by 280 (20 years $\times$ 14 corps = 280 groups), is as follows:

| Deaths | Predicted Number of Groups |
|--------|----------------------------|
| 0 | .497 × 280 = 139.2 |
| 1 | .348 × 280 =  97.4 |
| 2 | .122 × 280 =  34.2 |
| 3 | .028 × 280 =   7.8 |
| 4 | .005 × 280 =   1.4 |
|   | 280.0 |

How accurate is this description? The actual figures were as follows:

| Deaths | Actual Number of Groups |
|--------|-------------------------|
| 0 | 144 |
| 1 | 91 |
| 2 | 32 |
| 3 | 11 |
| 4 | 2 |

How close are the actual and predicted results? In Chapter 20, we will develop a measurement for "closeness of fit."

**Example 9.6**    During a recent leap year, an average of 3.26 new births per day occurred in Tompkins Community Hospital in Ithaca, New York. If the distribution were Poisson, the probabilities would be

| Births on 1 Day | Probability | |
|---|---|---|
| 0 | $e^{-3.26}$ | = .038 |
| 1 | $3.26e^{-3.26}$ | = .125 |
| 2 | $[(3.26)^2/2]e^{-3.26}$ | = .204 |
| 3 | $[(3.26)^3/3\ !]e^{-3.26}$ | = .222 |
| 4 | $[(3.26)^4/4\ !]e^{-3.26}$ | = .181 |
| 5 | $[(3.26)^5/5\ !]e^{-3.26}$ | = .118 |
| 6 | $[(3.26)^6/6\ !]e^{-3.26}$ | = .064 |
| 7 | $[(3.26)^7/7\ !]e^{-3.26}$ | = .030 |
| 8 | $[(3.26)^8/8\ !]e^{-3.26}$ | = .012 |
| 9 | $[(3.26)^9/9\ !]e^{-3.26}$ | = .004 |
| 10 | $[(3.26)^{10}/10\ !]e^{-3.26}$ | = .002 |
| | | 1.000 |

On the basis of these probabilities we have

| Births on 1 Day | Expected Number of Occurrences |
|---|---|
| 0 | .038 × 366 = 14 |
| 1 | .125 × 366 = 46 |
| 2 | .204 × 366 = 75 |
| 3 | .222 × 366 = 81 |
| 4 | .181 × 366 = 66 |
| 5 | .118 × 366 = 43 |
| 6 | .064 × 366 = 23 |
| 7 | .030 × 366 = 11 |
| 8 | .012 × 366 = 4 |
| 9 | .004 × 366 = 2 |
| 10 | .002 × 366 = 1 |

How accurate is this description? The actual figures obtained from working through hospital records are as follows:

| Births on 1 day | Actual Number of Occurrences |
|---|---|
| 0 | 15 |
| 1 | 42 |
| 2 | 74 |
| 3 | 95 |
| 4 | 53 |
| 5 | 44 |
| 6 | 23 |
| 7 | 13 |
| 8 | 6 |
| 9 | 1 |
| 10 | 0 |

**Example 9.7**  The Poisson is used in the following type of hypothesis test. Suppose that observations taken before a company began to dump pollutants into a river indicated that an average of three trout per hour swim past the dump site. An inspector plans to spend an hour at the site, and will issue a warning if she sees fewer than three trout.

a. If the pollutants have no effect on the trout, so the mean per hour is still three, what is the probability that a warning will be issued?

*Answer:*

$$P(\text{less than 3}) = P(0) + P(1) + P(2) = e^{-3} + 3e^{-3} + [3^2/2]e^{-3} = .423$$

b. If the pollutants kill half of the trout, what is the probability that a warning will or will not result?

*Answer:* In this case, the mean is .5(3) = 1.5, and thus we have

$$P(\text{warning}) = e^{-1.5} + 1.5e^{-1.5} + [(1.5)^2/2]e^{-1.5} = .809$$

$$P(\text{no warning}) = 1 - .809 = .191$$

Note that, with this inspection procedure, there still might be a warning (.423 probability) even if the pollutant is harmless, and there might be no warning (.191 probability) even if the pollutant kills half the fish. Different testing procedures will give different probabilities. Many factors go into determining what testing procedure is used, and mathematical calculations such as those shown here are among the most important.

# POISSON APPROXIMATION TO THE BINOMIAL

In this chapter we used the binomial to derive the formula for the Poisson distribution. Alternatively, an important application of the Poisson is as an *approximation to the binomial.* The Poisson approximation is of course valuable when $p$ and $n$ are not known, but the mean $np$ is known. However, the Poisson approximation is useful even when $p$ and $n$ are known, because the calculation of a Poisson probability is simpler than the calculation of the corresponding binomial probability. A rough rule of thumb is that the resulting approximation is "close" provided that

$$n \geq 20 \quad \text{and} \quad p \leq .05$$

Of course, how close is close enough, will depend on the degree of accuracy required in the particular application.

<u>**Example 9.8**</u>    In a binomial distribution with $p = .02$ and $n = 70$, what is the actual probability of three successes, and what is the Poisson approximation?

*Answer:* We calculate this probability in two ways:

Binomial: The probability of success is .02, so the probability of failure is .98. Thus we have

$$P(3 \text{ successes}) = \frac{70!}{67! \, 3!} (.02)^3 (.98)^{67} = .1131$$

Poisson: The mean is $70(.02) = 1.4$, and so

$$P(3 \text{ successes}) = \frac{(1.4)^3}{3!} e^{-1.4} = .1128$$

Note how much easier the Poisson calculation is than the binomial calculation.

<u>**Example 9.9**</u>    If 4% of the population is color-blind, what is the probability that in a group of 30 people, exactly one will be color-blind? At least one will be color-blind?

*Answer:* To use the Poisson approximation to this binomial, we first calculate the mean to be $30(.04) = 1.2$. Then

$$P(1 \text{ success}) = 1.2e^{-1.2} = .361$$

$$P(\text{at least 1 success}) = 1 - P(0 \text{ successes}) = 1 - e^{-1.2} = .699$$

<u>**Example 9.10**</u>    A delivery of ten items is received from a manufacturing plant at which 5% of the items produced are known to be defective. What is the probability of no defective items among the ten items? Of exactly one defective item?

*Answer:* Since $n = 10$ is not large, we do not expect to obtain a very good approximation to the normal using the Poisson (however, the result is surprisingly close). The mean is $10(.05) = 0.5$. Then

$$P(0 \text{ defectives}) = e^{-0.5} = .607 \quad (\text{Actual answer is } .599.)$$

$$P(1 \text{ defective}) = 0.5e^{-0.5} = .303 \quad (\text{Actual answer is } .315.)$$

# HISTOGRAMS

Finally, we obtain further insight into the shape of a Poisson distribution by looking at histograms for several different values of the mean $\mu$.

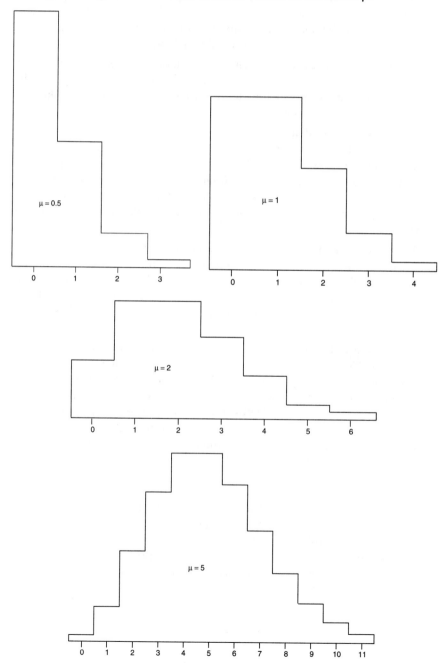

# EXERCISES

In each of the following exercises, assume that the distribution is Poisson.

1. According to the *Los Angeles Times,* during the past 5 years 1.6 elephant trainers, on average, were killed on the job in the United States every year. What is the probability of no death of an elephant trainer in 1 year? Of no deaths in a 2-year period? Of no deaths in 5 years?

2. *The New York Times* reports that large meteorites strike our atmosphere with the intensity of atomic bombs an average of eight times a year. What is the probability of no such meteorite strike in 1 year? Of five strikes in 1 year? Of eight in 1 year?

3. Angry executives of a company claim that the telephone switchboard operator makes an average of two wrong connections every hour, while the operator claims that he makes only one per hour. The manager of the secretarial staff decides to run a 2-hour test. If the operator makes four or more wrong connections in the 2 hours, he will be replaced. What is the probability that he will be replaced if the executives are correct? If he is correct?

4. A trucking company plans to purchase a new tire for its trucks. The owners decide to run a preliminary test using the new tires on a small number of their fleet of trucks. If there are no more than three flat tires in an initial 100,000 miles, the new tire will be accepted. What is the probability of acceptance if the tires actually average one flat per 50,000 miles? If they average one flat per 10,000 miles?

5. There were 231 deaths at an upstate New York hospital during a recent leap year. This is a binomial, but $n$ is large and unknown while $p$ is small and unknown, so we can use the Poisson as an approximation. Mathematically, on how many days would we have expected no deaths? One death? Two deaths? Three deaths? Four deaths? (The actual numbers were surprisingly close: 201 days with no deaths, 118 days with one death, 33 days with two deaths, 10 days with three deaths, 3 days with 4 deaths, and 1 day with five deaths.)

## Study Question

Regarding the Poisson as a limiting case of the binomial, determine the variance of a Poisson distribution with mean $\mu$. [*Hint:* What is the variance of a binomial distribution in terms of $p$, the probability of success, and $n$, the number of cases? What does the variance approach as $n$ becomes large and $p$ becomes small? What is the answer in terms of $\mu = np$?]

# 10
# NORMAL DISTRIBUTIONS

Some of the most useful probability distributions are symmetrical and bell-shaped:

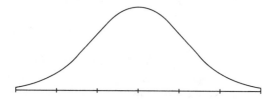

In this chapter we study one such distribution called the *normal distribution*. The normal distribution is valuable in describing various natural phenomena, especially those involving growth or decay. However, the real importance of the normal distribution in statistics is that it can be used to describe the results of sampling procedures.

The normal distribution, like the Poisson, can be viewed as a limiting case of the binomial. More specifically, we start with any fixed probability of success $p$, and consider what happens as $n$ becomes arbitrarily large. To obtain a visual representation of the limit, we draw histograms using an increasingly large $n$ for each fixed $p$.

Histograms for $p = .5$ and for $n = 4$, 10, and 25, respectively.

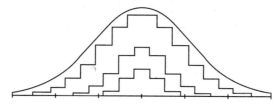

Histograms for $p = .1$ and for $n = 4$, 10, and 25, respectively.

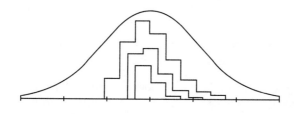

Histograms for $p = .7$ and for $n = 4$, 10, and 25, respectively.

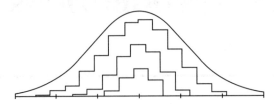

For greater clarity, different scales have been used for the histograms associated with different $n$'s. From the diagrams above it is reasonable to accept that as $n$ becomes larger without bound, the resulting histograms approach a smooth, bell-shaped curve. This is the curve associated with the *normal distribution*.

The normal curve is bell-shaped and symmetrical with an infinite base. Long, flat-looking tails cover many values but only a small proportion of the area. The flat appearance of the tails is deceptive. Actually, far out in the tails, the curve is dropping proportionately at an ever-increasing rate. In other words, when two intervals of equal length are compared, the one closer to the center may experience a greater numerical drop, but the one further out in the tail experiences a greater drop when measured as a proportion of the height at the beginning of the interval.

The mean here is the same as the median and is located at the center. We want a unit of measurement that will apply equally well to any normal distribution, and we choose a unit that arises naturally out of the curve's shape. There is a point on each side where the slope is steepest. These two points are called *points of inflection*, and the distance from the mean to either point is precisely equal to one standard deviation. Thus, it is convenient to measure distances under the normal curve in terms of $z$-scores (recall from Chapter 3 that $z$-scores are fractions or multiples of standard deviations from the mean).

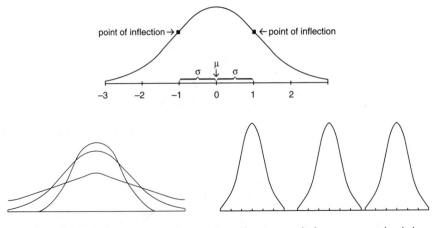

Normal curves with the same mean but different standard deviations

Normal curves with the same standard deviation but different means

Mathematically, the formula for the normal curve turns out to be $y = e^{-z^2/2}$ where $y$ is the *relative* height above a z-score. (By relative height we mean proportion of the height above the mean.) However, our interest is not so much in relative heights under the normal curve as in *proportionate areas*. The probability associated with any interval under the normal curve is equal to the proportionate *area* found under the curve and above the interval.

Table A in the Appendix gives proportionate areas under the normal curve. Because of the curve's symmetry, it is sufficient to give areas from the mean to positive $z$ values. Table A shows, for example, that between the mean and a z-score of 1.2 there is .3849 of the area:

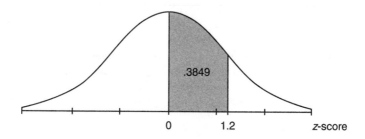

while .4834 of the area lies between the mean and a z-score of −2.13.

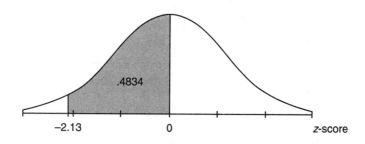

**Example 10.1**  The life expectancy of a particular brand of light bulbs is normally distributed with a mean of 1500 hours and a standard deviation of 75 hours.

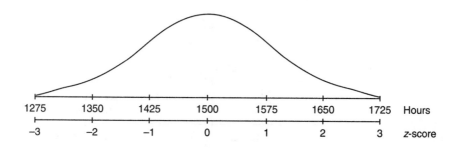

a. What is the probability that a bulb will last between 1500 and 1650 hours?

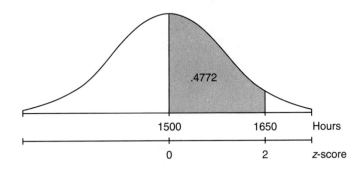

**Answer:** The $z$-score of 1500 is 0, the $z$-score of 1650 is (1650-1500)/75 = 2, and 2 in Table A gives a probability of .4772.

b. What percentage of the light bulbs will last between 1485 and 1500 hours?

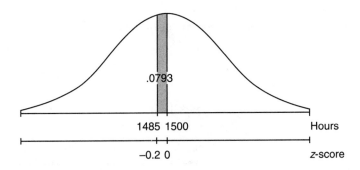

**Answer:** The $z$-score of 1485 is (1485-1500)/75 = −0.2, and 0.2 in Table A gives a probability of .0793 or 7.93%.

c. What is the probability that a bulb will last between 1416 and 1677 hours?

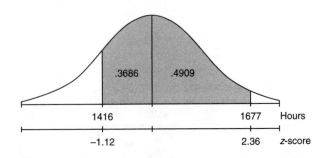

***Answer:*** The *z*-score of 1416 is (1416 − 1500)/75 = −1.12, and the *z*-score of 1677 is (1677 − 1500)/75 = 2.36. In Table A, 1.12 gives a probability of .3686, and 2.36 gives a probability of .4909. The total probability is .3686 + .4909 = .8595.

d. What is the probability that a light bulb will last between 1563 and 1648 hours?

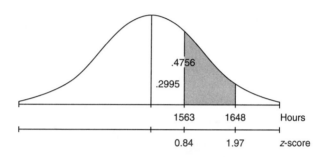

***Answer:*** The *z*-score of 1563 is (1563 − 1500)/75 = 0.84, and of 1648 is (1648 − 1500)/75 = 1.97. In Table A, 0.84 and 1.97 give probabilities of .2995 and .4756, respectively. Thus, between 1563 and 1648 there is a probability of .4756 − .2995 = .1761.

e. What is the probability that a light bulb will last less than 1410 hours?

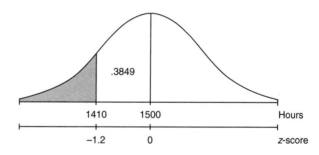

***Answer:*** The *z*-score of 1410 is (1410 − 1500)/75 = −1.2. In Table A, 1.2 gives a probability of .3849. The probability of being less than 1500 is .5, and so the probability of being less than 1410 is .5 − .3849 = .1151.

**Example 10.2** A packing machine is set to fill a cardboard box with a mean average of 16.1 ounces of cereal. Suppose the amounts per box form a normal distribution with standard deviation equal to 0.04 ounce.

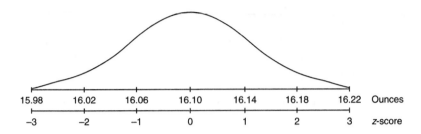

a. What percent of the boxes will end up with at least 1 pound of cereal?

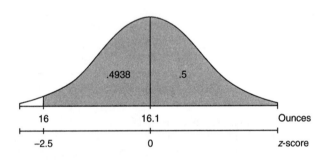

**Answer:** The *z*-score of 16 is (16 − 16.1)/0.04 = −2.5, and 2.5 in Table A gives a probability of .4938. Thus, the probability of more than 1 pound is .4938 + .5 = .9938 or 99.38%.

b. Ten percent of the boxes will contain more than what number of ounces?

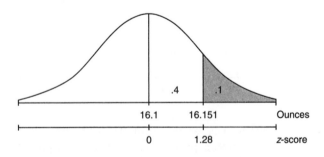

**Answer:** In Table A, we note that .4 area (actually, .3977) is found between the mean and a *z*-score of 1.28, so to the right of a 1.28 *z*-score must be 10% of the area. Converting the *z*-score of 1.28 into a raw score yields 16.1 + 1.28(0.04) = 16.151 ounces.

c. Eighty percent of the boxes will contain more than what number of ounces?

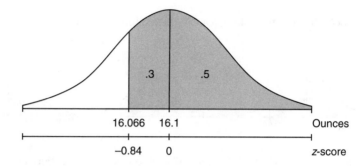

**Answer:** In Table A, we note that .3 area (actually, .2995) is found between the mean and a z-score of 0.84, so to the right of a −0.84 z-score must be 80% of the area. Converting the z-score of −0.84 into a raw score yields 16.1 − 0.84(0.04) = 16.066 ounces.

d. The middle 90% of the boxes will be between what two weights?

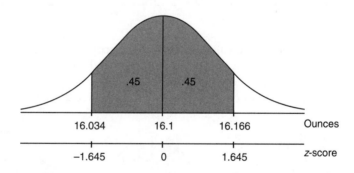

**Answer:** In Table A, we note that .4495 area is found between the mean and a z-score of 1.64, while a .4505 area is found between the mean and a z-score of 1.65, so use the z-score of 1.645. Then 90% of the area is between z-scores of −1.645 and 1.645. Converting z-scores to raw scores yields 16.1 − 1.645(0.04) = 16.034 ounces, and 16.1 + 1.645(0.04) = 16.166 ounces, for the two weights between which the middle 90% of the boxes will be.

# COMMONLY USED PROBABILITIES AND Z-SCORES

As can be seen from Example 10.2d, there is often an interest in the limits enclosing some specified middle percentage of the data. For future reference, the most frequently asked for of these limits are noted below in terms of z-scores.

Since .45 + .45 = .90, then 90% of the values are between z-scores of −1.645 and +1.645. Since .475 + .475 = .95, then 95% of the values are between z-scores of −1.96 and +1.96. Since .495 + .495 = .99, then 99% of the values are between z-scores of −2.58 and +2.58.

Sometimes the interest is in values with particular percentile rankings. For example:

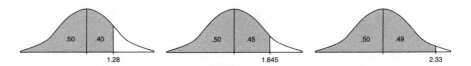

Ninety percent of the values are below a z-score of 1.28, ninety-five percent of the values are below a z-score of 1.645, and ninety-nine percent of the values are below a z-score of 2.33.

There are corresponding conclusions for negative z-scores:

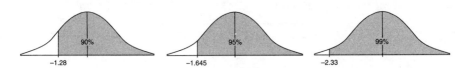

Ninety percent of the values are above a z-score of −1.28, ninety-five percent of the values are above a z-score of −1.645, and ninety-nine percent of the values are above a z-score of −2.33.

It is also useful to note the percentages corresponding to values falling between integer z-scores. For example:

Since .3414 + .3413 = .6826, then 68.26% of the values are between z-scores of −1 and +1. Since .4772 + .4772 = .9544, then 95.44% of the values are between *z*-scores of −2 and +2. Since .4987 + .4987 = .9974, then 99.74% of the values are between *z*-scores of −3 and +3.

**Example 10.3**  Suppose that the average height of adult males in a particular locality is 70 inches with a standard deviation of 2.5 inches.

a. If the distribution is normal, the middle 95% of males are between what two heights?

***Answer:*** As noted above, the critical *z*-scores in this case are ±1.96, so the two limiting heights are 1.96 standard deviations from the mean. Therefore, 70 ± 1.96(2.5) = 70 ± 4.9, or from 65.1 to 74.9 inches.

b. Ninety percent of the heights are below what value?

***Answer:*** The critical *z*-score is 1.28, so the value in question is 70 + 1.28(2.5) = 70 + 3.2 = 73.2 inches.

c. Ninety-nine percent of the heights are above what value?

***Answer:*** The critical *z*-score is −2.33, so the value in question is 70 − 2.33(2.5) = 70 − 5.825 = 64.175 inches.

d. What percentage of the heights are between *z*-scores of ±1? Of ±2? Of ±3?

***Answer:*** 68.26%, 95.44%, and 99.74%, respectively.

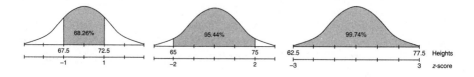

# FINDING MEANS AND STANDARD DEVIATIONS

If we know that a distribution is normal, we can calculate the mean $\mu$ and the standard deviation $\sigma$, using percentage information from the population.

**Example 10.4**  Given a normal distribution with a mean of 25, what is the standard deviation if 18% of the values are above 29?

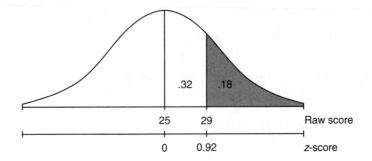

25    29        Raw score

0     0.92       z-score

***Answer:*** Looking for a .32 probability in Table A, we note that the corresponding *z*-score is 0.92. Thus 29 − 25 = 4 is equal to a standard deviation of 0.92, that is, 0.92σ = 4, and σ = 4/0.92 = 4.35.

**Example 10.5**  Given a normal distribution with a standard deviation of 10, what is the mean if 21% of the values are below 50?

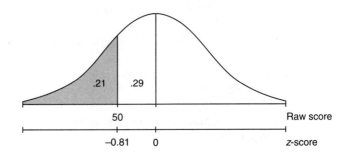

50        Raw score

−0.81    0    z-score

***Answer:*** Looking for a .29 probability in Table A leads us to a *z*-score of −0.81. Thus 50 is −0.81 standard deviation from the mean, so μ = 50 + 0.81(10) = 58.1.

**Example 10.6**  Given a normal distribution with 80% of the values above 125 and 90% of the values above 110, what are the mean and the standard deviation?

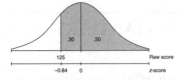 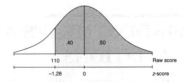

***Answer:*** Table A gives critical *z*-scores of −0.84 and −1.28. Thus we have (125 − μ)/σ = −0.84 and (110 − μ)/σ = −1.28. Solving the system {125 − μ = −0.84σ, 110 − μ = −1.28σ} simultaneously gives μ = 153.64 and σ = 34.09.

# NORMAL APPROXIMATION TO THE BINOMIAL

Many practical applications of the binomial involve examples where $n$ is large. However, for large $n$, binomial probabilities can be quite tedious to calculate. Since the normal can be viewed as a limiting case of the binomial, it is natural to use the normal to approximate the binomial in appropriate situations.

The binomial takes values only at integers, while the normal is continuous with probabilities corresponding to areas over intervals. Therefore, we must establish some technique for converting from one distribution to the other. For approximation purposes we do as follows. Each binomial probability will correspond to the normal probability over a unit interval centered at the desired value. Thus, for example, to approximate the binomial probability of eight successes we determine the normal probability of being between 7.5 and 8.5.

**Example 10.7**  Suppose that 15% of the cars coming out of an assembly plant have some defect. In a delivery of 40 cars what is the probability that exactly five cars have defects?

*Answer:* The actual answer is $\dfrac{40\,!}{35\,!\,5\,!}\,(.15)^5(.85)^{35}$, but clearly this involves a nontrivial calculation. To approximate the answer using the normal, we first calculate the mean $\mu$ and the standard deviation $\sigma$ as follows:

$$\mu = np = 40(.15) = 6$$

$$\sigma = \sqrt{np(1-p)} = \sqrt{40(.15)(.85)} = 2.258$$

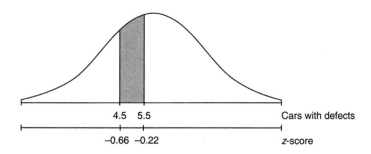

We then calculate the appropriate $z$-scores: $(4.5 - 6)/2.258 = -0.66$ and $(5.5 - 6)/2.258 = -0.22$. Looking up the corresponding probabilities in Table A, we obtain a final answer of $.2454 - .0871 = .1583$. (The actual answer is $.1692$.)

Even more useful are approximations relating to probabilities over intervals.

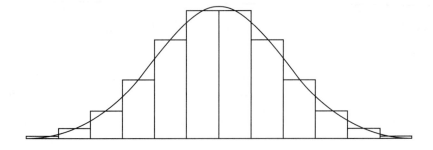

**Example 10.8**    If 60% of the population support massive federal budget cuts, what is the probability that in a survey of 250 people at most 155 people support such cuts?

**Answer:** The actual answer is the sum of 156 binomial expressions:

$$(.4)^{250} + \ldots + \frac{250!}{155!\,95!}\,(.6)^{155}(.4)^{95}$$

However a good approximation can be obtained quickly and easily by using the normal. We calculate $\mu$ and $\sigma$:

$$\mu = np = 250(.6) = 150$$

$$\sigma = \sqrt{np(1 - p)} = \sqrt{250(.6)(.4)} = 7.746$$

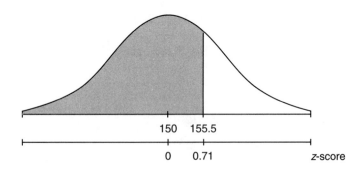

The binomial of at most 155 successes corresponds to the normal probability of $\leq 155.5$. The z-score of 155.5 is $(155.5 - 150)/7.746 = 0.71$. Using Table A leads us to a final answer of $.5000 + .2611 = .7611$.

Is the normal a "good" approximation? The answer, of course, depends on the error tolerances in particular situations. A general rule of thumb is that the normal is a good approximation to the binomial whenever both $np$ and $n(1 - p)$ are greater than 5.

**Example 10.9**   A particular form of cancer is fatal within 1 year in 30% of all diagnosed cases. A new drug is tried on 200 patients with this disease. Researchers will judge the new medication effective if at least 150 of the patients survive for longer than 1 year. If the medication has no effect, what is the probability that at least 150 patients survive for longer than 1 year?

*Answer:* The actual answer is the sum of 51 binomial expressions:

$$\frac{200!}{150!\,50!}(.7)^{150}(.3)^{50} + \frac{200!}{151!\,49!}(.7)^{151}(.3)^{49} + \ldots + (.7)^{200}$$

which would be very tedious to calculate. However, an approximate answer using the normal is readily calculated.

$$\mu = np = 200(.7) = 140$$

$$\sigma = \sqrt{np(1-p)} = \sqrt{200(.7)(.3)} = 6.48$$

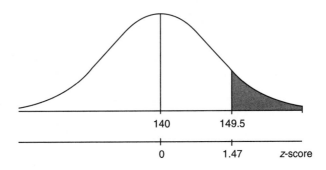

The binomial probability of 150 successes corresponds to the normal probability on the interval from 149.5 to 150.5, and thus the binomial probability of at least 150 successes corresponds to the normal probability of ≥149.5. The z-score of 149.5 is (149.5 − 140)/6.48 = 1.47. Using Table A gives us a final answer of .5000 − .4292 = .0708.

   In some situations both the normal and the Poisson are reasonable approximations to the binomial.

**Example 10.10**   Since airline companies know that 4% of all reservations received will be no-shows, they overbook accordingly. Suppose there are 126 seats on a plane, and the airline books 130 reservations. What is the probability that more than 126 confirmed passengers will show up? In other words, what is the probability that the number of no-shows will be three or less?

**Answer:** Since both $np = 130(.04) = 5.2$ and $n(1 - p) = 130(.96)$ $= 124.8$ are greater than 5, it is reasonable to determine a normal approximation. Also, both $p = .04 \le .05$ *and* $n = 130 \ge 20$ so it is reasonable to calculate a Poisson approximation.

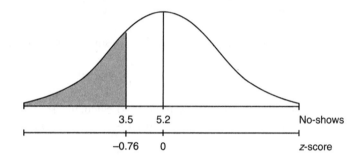

The normal approximation is calculated using $n = 130$ and $p = .04$ to derive $\mu = 5.2$ and $\sigma = 2.2343$. The $z$-score of 3.5 is $(3.5 - 5.2)/2.2343 = -0.76$, and the probability approximation is $.5 - .2764 = .2236$.

The Poisson approximation to the binomial is calculated as follows:

$$\frac{(5.2)^3}{3!} e^{-5.2} + \frac{(5.2)^2}{2} e^{-5.2} + (5.2)e^{-5.2} + e^{-5.2} = .2381$$

The actual answer is the sum of four binomial probabilities:

$$\frac{130!}{127!\,3!} (.04)^3(.96)^{127} + \frac{130!}{128!\,2!} (.04)^2(.96)^{128}$$

$$+130(.04)(.96)^{129} + (.96)^{130} = .2323$$

# EXERCISES

For exercises 1–10 assume that the distribution is normal.

1. A trucking firm determines that its fleet of trucks averages a mean of 12.4 miles per gallon with a standard deviation of 1.2 mpg on cross-country hauls. What is the probability that one of the trucks averages more than 13 mpg? Fewer than 10 mpg?

2. A factory dumps an average of 2.43 tons of pollutants into a river every week. If the standard deviation is 0.88 tons, what is the probability that in a week more than 3 tons are dumped? More than 1 ton? Fewer than 2 tons? Fewer than 5 tons?

3. An electronic product takes an average of 3.4 hours to move through an assembly line. If the standard deviation is 0.5 hours, what is the probability that an item takes between 3 and 4 hours? Between 2 and 3 hours? Between 4 and 4.4 hours?

4. The mean score on a college placement exam is 500 with a standard deviation of 100. What percentage of the students score above 700? Ninety-five percent score above what? Fifty percent score below what?

5. The average noise level in a restaurant is 30 decibels with a standard deviation of 4 decibels. Ninety-nine percent of the time the noise level is above what value? Ninety-five percent of the time it is below what value?

6. The mean income per household in a certain state is $9500 with a standard deviation of $1750. Ten percent of the household incomes are below what value? The middle ninety-five percent of the incomes are between what two values? The middle ninety-eight percent are between what two levels?

7. Jay Olshansky from the University of Chicago was quoted in *Chance News* as arguing that for the average life expectancy to reach 100, 18% of people would have to live to 120. What standard deviation must he be assuming for this statement to make sense?

8. Cucumbers grown on a certain farm have weights with a standard deviation of 2 ounces. What is the mean weight if 85% of the cucumbers weigh less than 16 ounces?

9. If 75% of all families spend more than $75 weekly for food, while 15% spend more than $150, what is the mean weekly expenditure and what is the standard deviation?

10. A coffee machine can be adjusted to deliver any fixed number of ounces of coffee. If the machine has a standard deviation in delivery equal to 0.4 ounce, what should be the mean setting so that an 8-ounce cup will overflow only 0.5% of the time?

11. The probability that a certain brand of painkiller will cure a simple headache is .81. If the medication is given to 29 individuals with simple headaches, what is the probability that fewer than 20 are cured? That exactly 24 are cured? That at least 24 are cured? Solve using the normal as an approximation to the binomial.

12. Assume that a baseball team has an "average" pitcher, that is, one whose probability of winning any decision is .5. If this pitcher has 30 decisions in a season, what is the probability he wins at least 20 games? Solve using the normal as an approximation to the binomial.

13. Five percent of all contract deliberations with a particular union result in strikes. In 50 such deliberations, what is the probability of exactly two strikes? Exactly three strikes? Compare the binomial solution with the Poisson and normal approximations.

## Study Questions

1. Earlier, the formula for the normal curve was given as $y = e^{-z^2/2}$, where $y$ is a relative height corresponding to the proportion of the value above the mean.

   a. Explain the significance that $z$ appears as a squared term (What does this say about symmetry?)

   b. Explain the significance of the *minus* sign (What happens to $y$ as $z$ becomes very large?)

   If we substitute $(x - \mu)/\sigma$ for $z$ and give actual rather than relative values for $y$, the formula becomes:

   $$y = \frac{1}{\sigma\sqrt{2\pi}} e^{-(x-\mu)^2/2\sigma^2}$$

   c. In this form, symmetry is around what value for $x$?

   d. What happens as $x$ becomes very large?

   e. What happens to the value of $y$ when $x = \mu$ if $\sigma$ is chosen to be larger, and what does this indicate about the graph? (You might also try to reason what the effect of the $\sigma^2$ term is on the graph if $\sigma$ is chosen to be larger.)

   You should be reaching a conclusion about how the value of $\sigma$ controls the relative extent to which the bell-shaped graph is peaked or spread out.

2. The normal is an example of a *continuous* probability distribution. More generally (for students with a little elementary calculus), if a function $f(x)$ is nonnegative on its domain, and if the definite integral of $f(x)$ over its domain is equal to 1, then $f(x)$ is called a *probability density* function. Verify that each of the following is a density function over the domain indicated:

   (1) $f(x) = \frac{3}{8}x^2$ *over* $0 \le x \le 2$     (2) $f(x) = \frac{1}{4}$ *over* $1 \le x \le 5$

   (3) $f(x) = \frac{1}{x^2}$ *over* $1 \le x < \infty$     (4) $f(x) = 3e^{-3x}$ *over* $0 \le x < \infty$

3. If $f(x)$ is a probability density function for a variable $x$, then the probability that $x$ falls between two values $a$ and $b$ is $\int_a^b f(x)dx$. Suppose that a telephone company has gathered data concerning the duration of calls lasting no more than 60 minutes, and has determined that $f(x) = (60 - x)/1800$ over $0 \le x \le 60$ (minutes) is the appropriate density function.

   (1) Find the probability that a randomly selected call lasts 3 minutes or less.

   (2) What proportion of calls last from 15 to 45 minutes?

   (3) What is the probability that a call will last more than 45 minutes?

# Part 4

# THE POPULATION MEAN

---

*The point of view of statistics is illustrated by the following example. Suppose a jar is filled with black and white balls. If we knew the exact number of balls of each color, we could apply probability theory to predict the likelihood of drawing a ball of a particular color or of drawing a sample of a particular mixture. On the other hand, suppose we do not know the composition of the whole jar, but we have drawn a sample. Statistics tells us with what confidence we can estimate the composition of the whole jar from that of the sample.*

*In our daily lives we frequently make judgments based on the characteristics of samples. A consumer looks at a few price tags and concludes that a clothing shop is a high- or low-priced store. A high school senior talks with a group of students and then forms an opinion about the college these students attend. A traveling salesperson rings doorbells for a day and then estimates the profitability of a new territory.*

*In all these examples complete observations are impossible or at least impracticable. The same is true in more structured situations. A medical researcher can't test everyone who has a particular form of cancer, a retailer can't test every flashbulb from a shipment (there would be none left to sell!), and a pollster can't survey every potential voter.*

*Thus we use samples to draw inferences about characteristics of the whole population. Many questions arise. With what size and in what manner should a sample be chosen? What conclusions about the population can be drawn from the sample? With what degree of confidence can these conclusions be stated? In answering these questions, we keep in mind that we are usually considering only a few members out of a large population and thus can never make any inference about the whole population with 100% certainty. We can, however, draw inferences with specified degrees of certainty.*

*In Chapters 11–15 we learn how samples are used both to establish confidence intervals of population means and to analyze hypothesis tests about population means. We consider single populations and the comparison of two or more populations. The effect of using large or small samples is also taken account of.*

# 11
# CONFIDENCE INTERVAL OF THE MEAN

Heights, weights, and like measurements tend to result in normal distributions, but normally distributed natural phenomena occur much less frequently than might be supposed. However, the normal curve has an importance in statistics that is independent of whether or not it appears in nature. What is significant is that the results of many types of sampling experiments can be analyzed using the normal curve. For example, there is no reason to suppose that the amounts of money that different people spend in grocery stores are normally distributed. However, if every day we survey 30 people leaving a store and determine the average grocery bill, these daily averages will have a nearly normal distribution. This statistical key will enable us to use a *sample mean* to estimate a *population mean.*

## DISTRIBUTION OF SAMPLE MEANS

We are interested in estimating the mean of a population. For our estimate we could simply randomly pick a single element of the population, but we then would have little confidence in our answer. Suppose instead we pick 100 elements and calculate their average. It is intuitively clear that the resulting sample mean has a greater chance of being closer to the mean of the whole population than does the value for any individual member of the population.

When we pick a sample and measure its mean, we are finding exactly one sample mean out of a whole universe of sample means. To judge the significance of a single sample mean, we must know how sample means vary. Consider the set of means from all possible samples of a specified size. It is both apparent and reasonable that the sample means will be clustered about the mean of the whole population; furthermore, these sample means will have a tighter clustering than do the elements of the original population. In fact, we might guess that the larger the chosen sample size, the tighter will be the clustering.

How do we calculate the variance of the set of sample means? Suppose the variance of the population is $\sigma_{pop}^2$, and we are interested in samples of size $n$. Sample means are obtained by first summing together $n$ elements and then dividing by $n$. In Chapter 2 we learned that a set of sums has a variance equal to the sum of the variances associated with the original sets. In our case

$$\sigma^2_{\text{sums}} = \sigma^2_{\text{pop}} + \ldots + \sigma^2_{\text{pop}} = n\sigma^2_{\text{pop}}$$

Also from Chapter 2 we remember that when each element of a set is divided by some constant, the new variance is the old one divided by the square of the constant. In our case the means are obtained by dividing the sums by $n$, so the variance of the sample means is obtained by dividing the variance of the sums by $n^2$. Thus if $\sigma_{\bar{x}}$ symbolizes the variance of the sample means, we find that:

$$\sigma^2_{\bar{x}} = \frac{\sigma^2_{\text{sums}}}{n^2} = \frac{n\sigma^2_{\text{pop}}}{n^2} = \frac{\sigma^2_{\text{pop}}}{n}$$

We note that the variance of the set of sample means varies directly as the variance of the original population and inversely as the size of the samples. In terms of standard deviations, we have

$$\sigma_{\bar{x}} = \frac{\sigma_{\text{pop}}}{\sqrt{n}}$$

# CENTRAL LIMIT THEOREM

With the above result in hand we state the following principle which forms the basis of much of what we do in this chapter and following ones. It is a simplified statement of the *central limit theorem* of statistics.

Start with a population with a given mean $\mu$ and standard deviation $\sigma$. Pick $n$ sufficiently large (at least 30), and take all samples of size $n$. Compute the mean of each of these samples. Then:

1. The set of all sample means will be approximately *normally* distributed.
2. The *mean* of the set of sample means will equal $\mu$, the mean of the population.
3. The *standard deviation*, $\sigma_{\bar{x}}$, of the set of sample mean will be approximately equal to $\sigma/\sqrt{n}$, that is, equal to the standard deviation of the whole population divided by the square root of the sample size.

**Example 11.1**    Suppose that the average outstanding credit card balance for young couples is $650 with a standard deviation of $420. If 100 couples are selected at random, what is the probability that the mean outstanding credit card balance exceeds $700?

**Answer:** The sample size is over 30, so by the central limit theorem the set of sample means is approximately normally distributed with mean 650 and standard deviation $420/\sqrt{100} = 42$.

With a $z$-score of $(700 - 650)/42 = 1.19$, the probability that the sample mean exceeds 700 is $.5000 - .3830 = .1170$.

# CONFIDENCE INTERVAL ESTIMATE OF THE POPULATION MEAN

Using a measurement from a sample, we will never be able to say *exactly* what the population mean is; rather we will always say we have a certain confidence that the population mean lies in a particular *interval*. The central limit theorem gives us the probability that a sample mean lies within a specified interval around the population mean, but this is precisely the same as saying that the population mean lies within a specified interval around the sample mean (the distance from Ithaca to Elmira is the same as the distance from Elmira to Ithaca).

**Example 11.2**   A bottling machine is operating with a standard deviation of 0.12 ounce. Suppose that in a sample of 36 bottles the machine inserted an average of 16.1 ounces into each bottle.

a. Estimate the mean number of ounces in all the bottles this machine fills. More specifically, give an interval in which we are 95% certain that the mean lies.

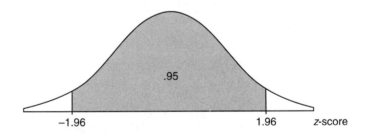

**Answer:** For samples of size 36, the sample means are approximately normally distributed with a standard deviation of $\sigma_{\bar{x}} = \sigma/\sqrt{n} = 0.12/\sqrt{36} = 0.02$.

From Chapter 10 we know that 95% of the sample means should be within 1.96 standard deviations of the population mean. Equivalently, we are 95% certain that the population mean is within 1.96 standard deviations of any sample mean.[1] In our case, $16.1 \pm 1.96(0.02) = 16.1 \pm 0.0392$, and we are 95% sure that the

---

[1] We cannot say there is a .95 *probability* that the population mean is within 1.96 standard deviations of a given sample mean. For a given sample mean, the population mean either is or is not within the specified interval, so the probability is either 1 or 0.

mean number of ounces in all bottles is between 16.0608 and 16.1392. This is called a *95% confidence interval estimate.*

b. How about a 99% confidence interval estimate?

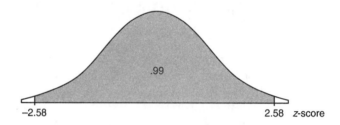

**Answer:** Here, $16.1 \pm 2.58(0.02) = 16.1 \pm 0.0516$, and we are 99% sure that the mean number of ounces in all bottles is between 16.0484 and 16.1516.

Note that, when we wanted a higher certainty (99% instead of 95%), we had to settle for a larger, less specific interval ($\pm0.0516$ instead of $\pm0.0392$).

From Example 11.2 we can give simple formulas for the 95% and 99% confidence interval estimates, respectively:

$$\bar{x} \pm 1.96\sigma_{\bar{x}} = \bar{x} \pm 1.96\frac{\sigma}{\sqrt{n}}$$

$$\bar{x} \pm 2.58\sigma_{\bar{x}} = \bar{x} \pm 2.58\frac{\sigma}{\sqrt{n}}$$

Confidence intervals other than 95% and 99% are obtained by replacing 1.96 or 2.58 by the appropriate $z$-score.

**Example 11.3** In a certain plant, batteries are being produced with a life expectancy that has a variance of 5.76 months squared. Suppose the mean life expectancy in a sample of 64 batteries is 12.35 months.

a. Find a 90% confidence interval estimate of life expectancy for all the batteries produced in this plant.

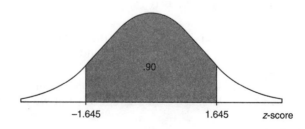

**Answer:** The standard deviation of the population is $\sigma = \sqrt{5.76} = 2.4$, and the standard deviation of the sample means is $\sigma_{\bar{x}} =$

$\sigma/\sqrt{n} = 2.4/\sqrt{64} = 0.3$. The 90% confidence interval estimate for the population mean is $12.35 \pm 1.645(0.3) = 12.35 \pm 0.4935$. Thus we are 90% certain that the mean life expectancy of the batteries is between 11.8565 and 12.8435 months.

b. What would the 90% confidence interval estimate be if the sample mean of 12.35 had come from a sample of 100 batteries?

***Answer:*** The standard deviation of sample means would then have been $\sigma_{\bar{x}} = \sigma/\sqrt{n} = 2.4/\sqrt{100} = 0.24$, and the 90% confidence interval estimate would be $12.35 \pm 1.645(0.24) = 12.35 \pm 0.3948$.

Note that when the sample size increased (from 64 to 100), the same sample mean resulted in a narrower, more specific interval ($\pm0.3948$ versus $\pm0.4935$).

**Example 11.4**    A new drug results in lowering heart rates by varying amounts with a standard deviation of 2.49 beats per minute.

a. Find a 95% confidence interval estimate for the mean lowering of heart beats in all patients if a 50 person sample averages a drop of 5.32 beats per minute.

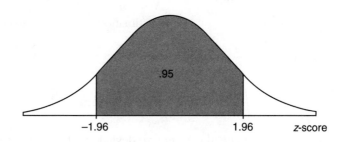

***Answer:*** The standard deviation of sample means is

$$\sigma_{\bar{x}} = \sigma/\sqrt{n} = 2.49/\sqrt{50} = 0.352.$$

We are 95% certain that the mean lowering of heart beats is in the range $5.32 \pm 1.96(0.352) = 5.32 \pm 0.69$ or between 4.63 and 6.01 heartbeats per minute.

b. With what certainty can we assert that the new drug lowers heart rates by a mean of $5.32 \pm 0.75$ beats per minute?

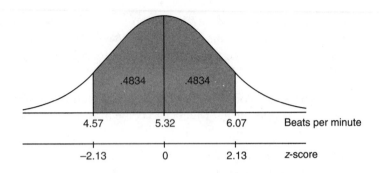

**Answer:** Converting 0.75 to a *z*-score yields 0.75/0.352 = 2.13. For a *z*-score of 2.13 Table A gives .4834, and our answer is 2(.4834) = .9668. In other words, 5.32 ± 0.75 beats per minute is a 96.68% confidence interval estimate of the mean lowering of heart rate effected by this drug.

# SUBSTITUTION OF *s* WHEN σ IS UNKNOWN

Frequently we do not know σ, the population standard deviation. In such cases, we must use *s*, the *standard deviation of the sample*, as an estimate for σ. However, if we compute *s* as $\sqrt{\dfrac{\Sigma(x - \bar{x})^2}{n}}$, the resulting *s* values from samples of size *n* tend to give underestimations for σ. Fortunately, there is a correction factor, $\sqrt{\dfrac{n}{n-1}}$. Noting that

$$\sqrt{\frac{n}{n-1}} \sqrt{\frac{\Sigma(x - \bar{x})^2}{n}} = \sqrt{\frac{\Sigma(x - \bar{x})^2}{n-1}} ,$$

we will always calculate the standard deviation of a sample by using the formula

$$s = \sqrt{\frac{\Sigma(x - \bar{x})^2}{n-1}}$$

whenever we intend to use *s* as an estimate for σ. For large sample size *n*, the correction becomes quite small. As there was for σ, there is an arithmetical tool for calculating *s*:

$$s = \sqrt{\frac{\Sigma(x - \bar{x})^2}{n-1}} = \sqrt{\frac{\Sigma x^2 - \dfrac{(\Sigma x)^2}{n}}{n-1}}$$

**Example 11.5**  An advertiser wishes to determine the mean number of hours per week that teenagers spend before television sets. The results of 500 interviews are as follows: $\Sigma x = 16{,}475$ and $\Sigma(x - \bar{x})^2 = 48{,}907$. Determine a 98% confidence interval estimate.

*Answer:* We first calculate the sample mean and standard deviation:

$$\bar{x} = \frac{\Sigma x}{n} = \frac{16475}{500} = 32.95$$

$$s = \sqrt{\frac{\Sigma(x - \bar{x})^2}{n - 1}} = \sqrt{\frac{48907}{499}} = 9.9$$

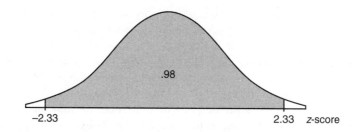

.98

−2.33                                   2.33   z-score

We use $s = 9.9$ as an estimate for the population standard deviation $\sigma$ to find the standard deviation of the sample means: $\sigma_{\bar{x}} = \sigma/\sqrt{n} = 9.9/\sqrt{500} = 0.443$. We can be 98% sure that the mean number of hours per week that teenagers spend before television sets is in the range $32.95 \pm 2.33(0.443) = 32.95 \pm 1.03$, or between 31.92 and 33.98 hours.

**Example 11.6**  A survey was conducted on 250 out of 125,000 families living in a city. The average amount of income tax paid per family in the sample was $3540 with a standard deviation of $1150. Establish a 99% confidence interval estimate for the total taxes paid by all the families in the city.

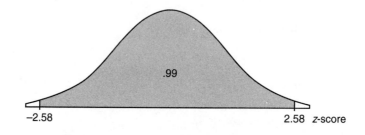

.99

−2.58                                   2.58  z-score

***Answer:*** We are given $\bar{x} = 3540$ and $s = 1150$. We use $s$ as an estimate for $\sigma$ and calculate the standard deviation of the sample means:

$$\sigma_{\bar{x}} = \sigma/\sqrt{n} = 1150/\sqrt{250} = 72.73$$

A 99% confidence interval estimate for the mean tax paid by each family is $3540 \pm 2.58(72.73) = 3540 \pm 187.64$. A 99% confidence interval estimate for the total taxes paid by all the families is $125,000(3540 \pm 187.64) = 442,500,000 \pm 23,455,000$ or between $419,045,000 and $465,955,000.

# SELECTING A SAMPLE SIZE

Statistical principles are useful not only in analyzing data, but also in setting up experiments. One consideration involves the choice of a *sample size*. In this chapter we have made interval estimates of population means and have seen that each inference must go hand in hand with an associated confidence level statement. Generally, if we want a smaller, more precise interval estimate, we either decrease the degree of confidence or increase the sample size. Similarly, if we want to increase the degree of confidence, we may either accept a wider interval estimate or increase the sample size. Thus, choosing a larger sample size seems always desirable; in the real world, however, time and cost considerations are involved.

**Example 11.7** Ball bearings are manufactured by a process that results in a standard deviation in diameter of 0.025 inch. What sample size should be chosen if we wish to be 99% sure of knowing the diameter to within ±0.01 inch?

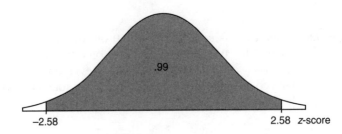

-2.58          .99          2.58 z-score

***Answer:*** We have

$$\sigma_{\bar{x}} = \sigma/\sqrt{n} = 0.025/\sqrt{n}$$

and $2.58\sigma_{\bar{x}} \leq 0.01$. Thus $2.58\,(0.025/\sqrt{n}) \leq 0.01$. Algebraically, we find that $\sqrt{n} \geq 2.58(0.025)/0.01 = 6.45$, so $n \geq 41.6$. We choose a sample size of 42.

A formula for the above calculation is

$$n = \left(\frac{z\sigma}{error}\right)^2$$

which in this problem gives $n = [(2.58)(0.025)/0.01]^2 = 41.6$.

**Example 11.8** A government investigator plans to test for the mean quantity of a particular pollutant that a manufacturer is dumping per day into a river. She needs an estimate that is within 50 grams at the 95% confidence level. If previous measurements indicate that the variance is approximately 21,800 grams$^2$, how many days should she include in the sample?

*Answer:* We have $\sigma^2 = 21,800$, so $\sigma = \sqrt{21,800} = 147.65$. Then $\sigma_{\bar{x}} = \sigma/\sqrt{n} = 147.65/\sqrt{n}$ and $1.96\sigma_{\bar{x}} \leq 50$. Thus $1.96(147.65/\sqrt{n}) \leq 50$, $\sqrt{n} \geq 1.96(147.65)/50 = 5.788$, and $n \geq 33.5$. The investigator should sample 34 days' dumping.

**Example 11.9** A sociologist is designing an experiment to determine the mean age of U.S. citizens who have strong opinions against committing funds for military operations in Bosnia. She has determined that, for a 90% confidence estimate of the mean age to within $\pm 3.0$ years, she will need to survey 100 individuals having the specified opinions. What would be the sample size for a 90% confidence estimate of the mean age to within $\pm 1.5$ years? To within $\pm 1.0$ year?

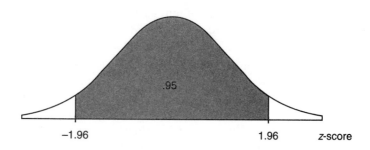

-1.96     .95     1.96    z-score

*Answers:* We have $\sigma_{\bar{x}} = \sigma/\sqrt{100}$ and $1.645\sigma_{\bar{x}} \leq 3.0$, or $1.645(\sigma/\sqrt{100}) \leq 3.0$. We want to know $n_1$ and $n_2$ where $1.645(\sigma/\sqrt{n_1}) \leq 1.5$ and $1.645(\sigma/\sqrt{n_2}) \leq 1.0$. Algebraically, we find that $\sqrt{n_1} \geq 1.645\sigma/1.5$, and $\sqrt{n_2} \geq 1.645\sigma/1.0$, and from the known sample size we have $1.645\sigma \leq 3.0\sqrt{100}$. Thus, $\sqrt{n_1} \geq 3.0\sqrt{100}/1.5$, so $n_1 \geq 400$. Also, $\sqrt{n_2} \geq 3.0\sqrt{100}/1.0$, so $n_2 \geq 900$.

Two points are worth noting. First, what would have been the result if, for example, 95% had been used instead of 90% for the confidence level?

**Answer:** Each 1.645 would have been replaced by 1.96, and instead of dividing 1.645 by 1.645, we would divide 1.96 by 1.96. The resulting answers, 400 and 900, however, would not have changed.

Second, how much must the sample size be increased in order to cut the interval estimate in half? To a third?

**Answer:** To cut the interval estimate in half (from $\pm 3.0$ to $\pm 1.5$), we would have to increase the sample size fourfold (from 100 to 400). To cut the interval estimate to a third (from $\pm 3.0$ to $\pm 1.0$), we would have to increase the sample size ninefold (from 100 to 900).

More generally, if we want to divide the interval estimate by $d$ without affecting the confidence level, we must increase the sample size by a multiple of $d^2$.

# EXERCISES

1. In a test for acid rain, 49 water samples showed a mean pH level of 4.4 with a standard deviation of 0.35. Find a 90% confidence interval estimate for the mean pH level.

2. One gallon of gasoline is put into each of 30 test autos, and the resulting mileage figures tabulated with $\bar{x} = 28.5$ and $s = 1.2$. Determine a 95% confidence interval estimate of the mean mileage.

3. For a sample of 490 men judged low in fitness, the resting heart rate (beats/min) had a mean of 64 with a standard deviation of 10; the triglyceride count (mg/dl) showed a mean of 126 with a standard deviation of 68; and the vital capacity (mL) had a mean of 4029 with a standard deviation of 860. The comparable figures among a sample of 487 men judged high in fitness were $\bar{x} = 59$ and $s = 9$ for heart rate, $\bar{x} = 101$ and $s = 48$ for triglycerides, and $= 4827$ and $s = 867$ for vital capacity (*New England Journal of Medicine*, February 25, 1993, page 535.)
   a. Find 90% confidence interval estimates for the mean heart rates of the men judged low and judged high in fitness.
   b. Find 95% confidence interval estimates for the triglyceride means of both samples.
   c. Find 99% confidence interval estimates for the vital capacity means of both samples.

4. The number of accidents per day in a large factory is noted for each of 64 days with $\bar{x} = 3.58$ and $s = 1.52$. With what degree of confidence can we assert that the mean number of accidents per day in the factory is between 3.20 and 3.96?

5. A company owns 335 trucks. For a sample of 30 of these trucks, the average yearly road tax paid is $9540 with a standard deviation of $1205. What would be a 99% confidence interval estimate for the total yearly road taxes paid for the 335 trucks?

6. What sample size should be chosen to find the mean number of absences per month by school children to within $\pm.2$ at a 95% confidence level if it is known that the standard deviation is 1.1?

7. Hospital administrators wish to learn the average length of stay of all surgical patients. A statistician determines that, for a 95% confidence level estimate of the average length of stay to within $\pm 0.5$ days, 50 surgical patients' records will have to be examined. How many records should be looked at for a 95% confidence level estimate to within $\pm 0.25$ days? $\pm 1$ day?

8. The 1995 profits (in millions of dollars) for a sample of 35 corporations are: 23, 43, 12, 3, 45, 41, 0, 23, 18, 37, $-12$, 15, 71, 22, 10, 33, 34, 61, 20, $-21$, 0, 29, 18, 57, 58, 0, 35, 38, 21, $-32$, $-39$, 17, 21, 40, 29. Establish a 98% confidence interval estimate for the 1995 profits of all corporations.

9. The ages of 49 company presidents give $\Sigma x = 2753.8$ and $\Sigma(x - \bar{x})^2 = 1348.32$. Determine a 90% confidence interval estimate for the mean age of company presidents.

10. In a sample of 36 hours of television programming, the number of minutes of commercials for each hour is noted and the totals tabulated as $\Sigma x = 378$ and $\Sigma x^2 = 4123.35$. Determine a 95% confidence interval estimate for the mean number of commercial minutes per hour of television programming.

## Study Questions

1. Consider a population of size $N = 5$ consisting of the elements {3, 6, 8, 10, 18}.
   a. Determine the population mean $\mu$ and the population standard deviation $\sigma$. (For use in Study Question 3, note that the population variance $\sigma^2 = 25.6$.)
   b. List all possible samples of size $n = 2$ [there are $C(5,2) = 10$ of them], and determine the mean $\bar{x}$ of each sample.
   c. Show that the mean of the set of ten sample means is equal to the population mean $\mu$.

d.  Show that the standard deviation of the set of ten sample means is equal to

$$\sigma_{\bar{x}} = \frac{\sigma}{\sqrt{n}} \sqrt{\frac{N-n}{N-1}}$$

[You should find that the calculation from the definition and from this formula are both equal to 3.0984.]

e.  More generally, what can be said about $\frac{\sigma}{\sqrt{n}} \sqrt{\frac{N-n}{N-1}}$ if the population size $N$ is very large?

2.  a.  For the same set as in Study Question 1, {3, 6, 8, 10, 18}, form all possible "ordered" samples of size 2. In other words, (3,6) and (6,3) will be considered different samples, and duplications such as (3,3) will also be allowed.

b.  Determine the mean for each of the $5 \times 5 = 25$ resulting samples.

c.  Show that the standard deviation of the set of these 25 sample means is precisely equal to $\sigma/\sqrt{n}$, where $\sigma$ is the population standard deviation and $n = 2$ is the sample size.

3.  Consider again the set of 25 "ordered" samples from Study Question 2.

a.  For each compute the variance using $s^2 = \Sigma(x - \bar{x})^2/n$. (Note that each $\Sigma$ is over only two terms, and $n = 2$.)

b.  Sum and divide by 25 to find the average of these variances.

c.  Repeat the calculations this time using $s^2 = \Sigma(x - \bar{x})^2/(n - 1)$.

Note that using $n - 1$ instead of $n$ results in the average of the sample variances being 25.6 which is precisely the variance $\sigma^2$ of the original population {3, 6, 8, 10, 18}. This example is true more generally and indicates why we define the sample standard deviation by

$$s = \sqrt{\frac{\Sigma(x - \bar{x})^2}{n - 1}} \qquad \textit{instead of} \qquad \sqrt{\frac{\Sigma(x - \bar{x})^2}{n}}$$

4.  In this chapter we noted the effect on sample size if the interval estimate is changed without changing the confidence level. What happens if we want to change the confidence level without changing the interval estimate? For example, for a given interval estimate, what is the effect on sample size if we go from a 90% confidence level to a 95% confidence level? From 90% to 99%?

# 12
# HYPOTHESIS TEST OF THE MEAN

Closely related to the problem of estimating a population mean is the problem of testing a hypothesis about a population mean. For example, a consumer protection agency might determine an interval estimate for the mean nicotine content of a particular brand of cigarettes, or, alternatively, it might test a manufacturer's claim about the mean nicotine content of his cigarettes. An agricultural researcher could find an interval estimate for the mean productivity gain caused by a specific fertilizer, or alternatively, she might test the developer's claimed mean productivity gain. A social scientist might ascertain an interval estimate for the mean income level of migrant farmers, or alternatively, he might test a farm bureau's claim about the mean income level. In each of these cases, the experimenter must decide whether the interest lies in an interval estimate of a population mean or in a hypothesis test of a claimed mean.

## NULL AND ALTERNATIVE HYPOTHESES

The general testing procedure is to choose a specific hypothesis to be tested, called the *null hypothesis*, pick an appropriate sample, and then use measurements from the sample to determine the likelihood of the null hypothesis. Just as in Chapter 11, conclusions are stated, not with absolute certainty, but rather with associated significance levels.

There are two types of possible errors we will consider: the error of mistakenly rejecting a true null hypothesis and the error of mistakenly failing to reject a false null hypothesis. For example, a manufacturer claims that the mean lifetime of an electronic component of his product is 1500 hours. A researcher believes that the true figure is lower and will test the 1500-hour claim by measuring the lifetime of each element in a sampling of components. The researcher decides that, if the sample average is less than 1450 hours, she will reject the manufacturer's claim. Alternatively, if the sample average exceeds 1450 hours, she will conclude that she does not have sufficient evidence to reject the 1500-hour claim.

The claim to be tested, the *null hypothesis*, denoted as $H_o$, is usually what we want to challenge, and is stated in terms of a specific value for a population parameter. In the case considered here,

$$H_o: \mu = 1500$$

The *alternative hypothesis*, denoted as $H_a$, is usually what we want to establish, and is stated in terms of an inequality such as $<$, $>$, or $\neq$. In this case

$$H_a: \mu < 1500$$

If the alternative hypothesis involves the inequality $<$ or $>$, there is one *critical value c* that separates the null hypothesis *rejection region* from the *fail to reject region*. In this case

$$c = 1450 \text{ and } \frac{\text{rejection region} \quad | \quad \text{fail to reject region}}{1450}$$

If the 1500-hour claim is true but the sample mean happens to be less than 1450, the researcher will *mistakenly reject* the null hypothesis. This is called a *Type I error*, and the probability of committing such an error is called the $\alpha$-*risk*. If the 1500-hour claim is false, but the sample mean happens to be over 1450, the researcher will *mistakenly fail to reject* the null hypothesis. This is called a *Type II error*, and the probability of committing such an error is called the $\beta$-*risk*. (The Greek letters used to designate Type I and Type II errors are lower-case alpha and beta, respectively.)

Null hypothesis ($\mu = 1500$)

|  |  | True | False |
|---|---|---|---|
| Sample mean | Less than 1450 | Type I error *Mistakenly reject manufacturer's claim* | Correct decision *Reject false claim* |
|  | More than 1450 | Correct decision *Do not reject true claim* | Type II error *Mistakenly fail to reject manufacturer's false claim* |

Several questions immediately come to mind. How should the critical value $c$ be chosen? How can the probabilities of Type I and Type II errors be calculated? What effect does sample size have on the whole procedure? These and other queries will be answered in the discussion and examples of this chapter.

# CRITICAL VALUES AND α-RISKS

For a given critical value, such as that in the example above, we can sketch the distribution of sample means around the claimed population mean, and then note the region corresponding to the probability of a Type I error or $\alpha$-risk.

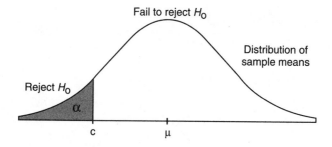

**Example 12.1** A coffee dispensing machine is supposed to drop 8 ounces of liquid into each paper cup, but a consumer believes that the actual amount is less. As a test he plans to obtain a sample of 36 cups of the dispensed liquid, and, if the mean content is less than 7.75 ounces, to reject the 8-ounce claim. If the machine operates with a standard deviation of 0.9 ounces, what is the α-risk?

***Answer:*** We have:

$$H_o: \mu = 8$$

$$H_a: \mu < 8$$

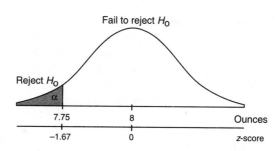

The standard deviation of sample means is $\sigma_{\bar{x}} = \sigma/\sqrt{n} = 0.9/\sqrt{36} = 0.15$. The z-score for 7.75 is $(7.75 - 8)/0.15 = -1.67$. Using Table A, we obtain $\alpha = .5000 - .4525 = .0475$. Thus, if the 8-ounce claim is correct, there is a .0475 probability that the consumer will still obtain a sample mean less than 7.75 and will mistakenly reject the claim.

**Example 12.2** A patient claims that he is consuming only 2000 calories per day, but a dietician suspects that the actual figure is larger. The dietician plans to check food intake on 30 days and will reject the patient's claim if the 30-day mean is more than 2100 calories. If the standard deviation (in calories per day) is 350, what is the probability of a Type I error?

***Answer:*** We have:

$$H_o: \mu = 2000$$

$$H_a: \mu > 2000$$

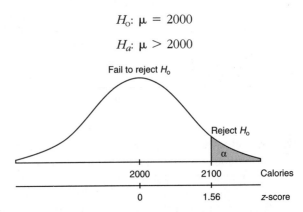

The standard deviation of sample means is $\sigma_{\bar{x}} = \sigma/\sqrt{n} = 350/\sqrt{30} = 63.9$. The $z$-score for 2100 is $(2100 - 2000)/63.9 = 1.56$, and from Table A we obtain $.5000 - .4406 = .0594$. Thus, if the 2000-calorie claim is correct, there is a .0594 probability that the dietician will still obtain a mean greater than 2100 and mistakenly reject the patient's claim.

**Example 12.3**   A government statistician claims that the mean monthly rainfall along the Liberian coast is 15.0 inches with a standard deviation of 12.0 inches. A meteorologist plans to test this claim with measurements over 3.5 years (42 months). If she finds a sample mean more than 2.0 inches different from the claimed 15.0 inches, she will reject the government statistician's claim. What is the probability that the meteorologist will mistakenly reject a true claim?

***Answer:*** We have:

$$H_o: \mu = 15.0$$

$$H_a: \mu \neq 15.0$$

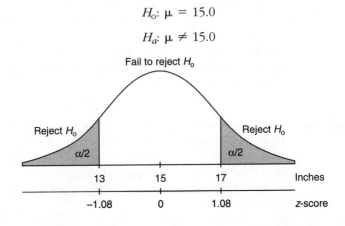

This is an example of a two-sided (or two-tailed) test, and the α-risk is computed by adding the probabilities corresponding to two regions. The standard deviation of the sample means is $\sigma_{\bar{x}}$ = 12.0/√42 = 1.85. The z-score for 17 is (17 − 15)/1.85 = 1.08; similarly, the z-score for 13 is −1.08. From Table A we obtain .5000 − .3599 = .1401, for a total probability of Type I error of .1401 + .1401 = .2802.

Examples 12.1–12.3 begin with critical values and proceed to find the corresponding α-risks. In most situations, however, this reasoning is reversed; that is, we choose acceptable α-risks and then calculate the corresponding critical values. The choice of α-risk is called the *significance level* of the test.

**Example 12.4**    A manufacturer claims that a particular automobile model will get 50 miles per gallon on the highway. The researchers at a consumer-oriented magazine believe that this claim is high and plan a test with a sample of 30 cars. Assuming the standard deviation between cars is 2.3 miles per gallon, determine the critical value for a test at the 5% significance level.

***Answer:*** We have:

$$H_o: \mu = 50$$

$$H_a: \mu < 50$$

$$\alpha = .05$$

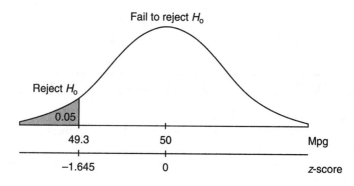

The standard deviation of sample means is $\sigma_{\bar{x}}$ = 2.3/√30 = 0.42. From Table A, the critical z-score corresponding to α = .05 is −1.645, which translates to a critical raw score of 50 − 1.645(0.42) = 49.3. Thus, if the sample mean is less than 49.3 miles per gallon, the manufacturer's claim should be rejected.

**Example 12.5**   A student handbook claims that college students study an average of 30 hours per week. A guidance counselor plans to test this claim at the 1% significance level. What are the critical values if the standard deviation in the number of hours that students study per week is 8 hours? Assume a sample of 100 students.

***Answer:*** We have:

$$H_o: \mu = 30$$

$$H_a: \mu \neq 30$$

$$\alpha = .01$$

$$\sigma_{\bar{x}} = 8/\sqrt{100} = 0.8$$

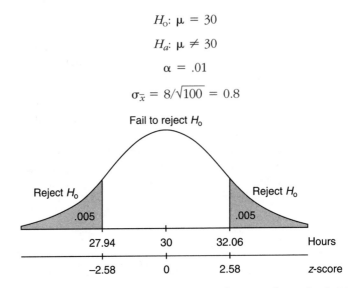

This is a two-sided (two-tailed) test, so for a total $\alpha$-risk of .01 we put a probability of $\frac{1}{2}$ (.01) = .005 on each side. A probability of .005 gives critical $z$-scores of $\pm2.58$. The corresponding critical hour values are $30 \pm 2.58(0.8) = 30 \pm 2.06$. Thus the handbook claim of 30 hours should be rejected if the sample mean is less than 27.94 or greater than 32.06 hours.

# DRAWING CONCLUSIONS; p-VALUES

Examples 12.1–12.5 illustrate the relationship between critical values and $\alpha$-risks. We now review the relevant terminology, and then give more complete examples of hypothesis testing.

The *null hypothesis* is a claim about a population that we usually hope to dispute, while the *alternative hypothesis* is an opposing claim that we hope to establish. In each of the above examples the null hypothesis is in the form of an equality statement about the population mean (such as $\mu = 25.1$), while the alternative hypothesis is stated as an inequality (such as $\mu < 25.1$ or $\mu > 25.1$ or $\mu \neq 25.1$).

We attempt to show that the null hypothesis is unacceptable by showing that it is improbable. Our testing procedure involves picking a sample and comparing the sample statistic (such as the mean $\bar{x}$) to the claimed population parameter (such as $\mu$). *Critical values* are chosen to gauge the significance of a sample statistic. If the sample statistic is far enough away from the claimed population parameter, we say that there is sufficient evidence to reject the null hypothesis. The probability of committing a *Type I error*, that is the $\alpha$-*risk*, is the probability of mistakenly rejecting a true null hypothesis. Later in this chapter we discuss more fully the *Type II error*, that of mistakenly failing to reject a false null hypothesis.

In working with critical values and $\alpha$-risks, we use the *central limit theorem*, that is that the set of all sample means is normally distributed around the population mean with a standard deviation equal to the population standard deviation divided by the square root of the sample size. In practice, the population standard deviation is usually unknown, and we use the sample standard deviation as an estimate for it.

The following examples illustrate the principles stated above and introduce the concept of *p-values*.

**Example 12.7**   An automotive company executive claims that a mean of 48.3 cars per dealership are being sold each month. A major stockholder believes this claim is high and runs a test by sampling 30 dealerships.

a. What conclusion is reached if the sample mean is 45.4 cars with a standard deviation of 15.4? Assume a 10% significance level.

***Answer:***

$$H_o: \mu = 48.3$$

$$H_a: \mu < 48.3$$

$$\alpha = .10$$

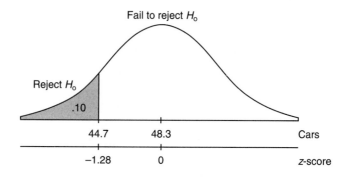

Here

$$\sigma_{\bar{x}} = \frac{\sigma}{\sqrt{n}} \approx \frac{s}{\sqrt{n}} = \frac{15.4}{\sqrt{30}} = 2.81.$$

With $\alpha = .10$, the critical $z$-score is $-1.28$. The critical number of cars is $48.3 - 1.28(2.81) = 44.7$. Since $45.4 > 44.7$, the stockholder should not reject the executive's claim.

b. If $\alpha$ had been chosen differently, would the executive's claim have been rejected?

***Answer:*** The sample statistic, 45.4, has a $z$-score of $(45.4-48.3)/2.81 = -1.03$. Using Table A, we note that if $\alpha$ was $.5000 - .3485 = .1515$ (or larger), the stockholder would have sufficient evidence to reject the 48.3-car claim.

With Example 12.7 as motivation, we define the *p-value* of a test to be the smallest value of $\alpha$ for which the null hypothesis would be rejected. Or, equivalently, if the null hypothesis is assumed to be true, the *p-value* of a sample statistic is the probability of obtaining a result as extreme as the one obtained. Note that the smaller the *p*-value, the more significant is the difference between the null hypothesis and the sample results. In Example 12.7, the *p*-value is .1515. Therefore, although the null hypothesis could not be rejected at the 10% (or even the 15%) significance level, it could be rejected at, for example, the 16% significance level.

**Example 12.8**    A manufacturer claims that a new brand of air-conditioning unit uses only 6.5 kilowatts of electricity per day. A consumer agency believes the true figure is higher and runs a test on a sample of size 50.

a. If the sample mean is 7.0 kilowatts with a standard deviation of 1.4, should the manufacturer's claim be rejected at a significance level of 5%? Of 1%?

***Answer:***

$$H_o: \mu = 6.5$$

$$H_a: \mu > 6.5$$

$$\alpha = .05$$

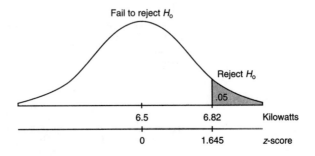

Fail to reject $H_o$

Reject $H_o$

.05

| 6.5 | 6.82 | Kilowatts |
| 0 | 1.645 | z-score |

Here

$$\sigma_{\bar{x}} = \frac{\sigma}{\sqrt{n}} \approx \frac{s}{\sqrt{n}} = \frac{1.4}{\sqrt{50}} = 0.198.$$

With $\alpha = .05$, the critical $z$-score is 1.645. There are actually two ways to proceed. We can convert 1.645 to a critical number of kilowatts and check whether 7.0 is greater than this critical number, or we can convert 7.0 to a $z$-score and compare this value to the critical $z$-score, 1.645. By the first method, 1.645 converts to a raw score of $6.5 + 1.645(0.198) = 6.83$; $7.0 > 6.83$, so reject $H_o$. By the second method, 7.0 converts to a $z$-score of $(7.0 - 6.5)/0.198 = 2.53$; $2.53 > 1.645$, so reject $H_o$. Thus, at the 5% significance level, the consumer agency should reject the manufacturer's claim.

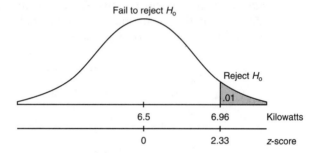

With $\alpha = .01$, the critical $z$-score is 2.33 with a corresponding critical number of kilowatts: $6.5 + 2.33(0.198) = 6.96$. Since $7.0 > 6.96$, the consumer should still reject the manufacturer's claim at the 1% significance level. Note that we could have compared 2.53, the $z$-score of 7.0, with 2.33 to reach the same conclusion.

b. What is the $p$-value of this test result?

***Answer:*** We have shown that the difference between the sample statistic ( = 7.0) and the claimed population parameter ($\mu = 6.5$) is significant at both the .05 and .01 levels. Calculation of the $p$-value gives a more complete picture of the significance of the observed difference.

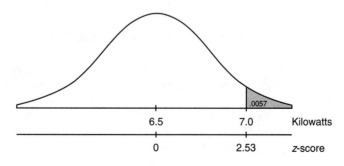

The $z$-score of 7.0 is 2.53; using Table A, we find the $p$-value to be $.5000 - .4943 = .0057$. Note that with this calculation we could have quickly concluded that at the .05 and .01 significance levels there is sufficient evidence to reject $H_o$. We also see that there would have been sufficient evidence at the .006 level, but not at the .005 level.

**Example 12.9**  A local chamber of commerce claims that the mean family income level is $12,250. An economist runs a hypothesis test, using a sample of 135 families, and finds a mean of $11,500 with a standard deviation of $3180. Should the $12,250 claim be rejected at a 5% level of significance?

*Answer:* Note that this is a two-sided test; that is, the economist suspects that the claim may be incorrect, but does not know whether it is probably too high or too low.

$$H_o: \mu = 12{,}250$$

$$H_a: \mu \neq 12{,}250$$

$$\alpha = .05$$

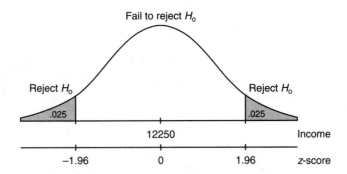

Here

$$\sigma_{\bar{x}} = \frac{\sigma}{\sqrt{n}} \approx \frac{s}{\sqrt{n}} = \frac{3180}{\sqrt{135}} = 273.7.$$

For a total $\alpha = .05$, we place a probability of $.5(.05) = .025$ on each side, and note the critical $z$-scores of $\pm 1.96$. These translate into critical incomes of $12{,}250 \pm 1.96(273.7)$ or $11{,}713.45$ and $12{,}786.45$. Since $11{,}500 < 11{,}713.45$, the chamber of commerce claim is rejected.

Alternatively, we could have changed 11,500 into a $z$-score: $(11{,}500 - 12{,}250)/273.7 = -2.74$ and observed that $-2.74$ is less than the critical $z$-score, $-1.96$.

We could have also reasoned in terms of the $p$-value: For the $z$-score, $-2.74$, Table A gives a probability of $.5000 - .4969 = .0031$. Since this is a two-tailed test, where disagreement with the null hypothesis can be in either of two directions, the $p$-value is twice the $.0031$ probability, or $.0062$. Since $.0062 < .05$, there is sufficient evidence to reject $H_o$.

**Example 12.10** A pharmaceutical company claims that a medication will produce a desired effect for a mean time of 58.4 minutes. A government researcher runs a hypothesis test of 250 patients and tabulates the following data with reference to the durations of effect in minutes: $\Sigma x = 14,875$ and $\Sigma(x-\bar{x})^2 = 17,155$. Should the company's claim be rejected at a significance level of 10%? Of 2%?

**Answer:**

$$\bar{x} = \frac{\Sigma x}{n} = \frac{14875}{250} = 59.5$$

$$s = \sqrt{\frac{\Sigma(x - \bar{x})^2}{n - 1}} = \sqrt{\frac{17155}{249}} = 8.3$$

$$\sigma_{\bar{x}} \approx \frac{s}{\sqrt{n}} = \frac{8.3}{\sqrt{250}} = 0.525$$

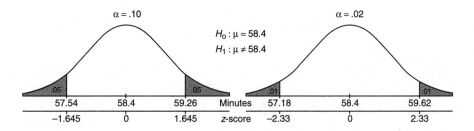

For $\alpha = .10$, the critical values (in minutes) are $58.4 \pm 1.645(0.525)$, or 57.54 and 59.26. For $\alpha = .02$, the critical values (in minutes) are $58.4 \pm 2.33(0.525)$, or 57.18 and 59.62. Now 59.5 > 59.26, but 59.5 < 59.62, so the researcher would reject the company's claim at the 10% significance level, but not at the 2% significance level. In other words, if she is willing to be wrong 10 times out of 100, she will reject the company's claim; but if she is willing to accept only 2 errors in 100 decisions, she will conclude that the evidence is not strong enough to reject the claim.

Again, the $p$-value can be used to measure specifically the strength of evidence for rejection of $H_o$. In this example, the $z$-score for 59.5 is $(59.5 - 58.4)/0.525 = 2.10$. Using Table A, we find the corresponding probability to be .5000 − .4821 = .0179. Doubling because the test is two-sided results in a $p$-value of 2(.0179) = .0358.

# TYPE II ERRORS AND β-RISKS

The preceding discussion and examples centered around critical values and corresponding Type I errors. The question arises, why not always choose the α-risk to be extremely small, such as .001 or .0001, and so virtually eliminate the possibility of mistakenly rejecting a correct null hypothesis? The problem is that this course of action would simultaneously increase the chance of never rejecting the null hypothesis, even if it were false.

Thus we are led to a discussion of the Type II error, that is, a mistaken failure to reject a false null hypothesis. As defined earlier, the probability of a Type II error is called the β-*risk*. We shall see that there is a different value of β for each possible correct value for the population parameter.

|  |  | Null hypothesis | |
|---|---|---|---|
|  |  | True | False |
| Decision to | Accept $H_o$ | Correct decision | Type II error |
|  | Reject $H_o$ | Type I error | Correct decision |

**Example 12.11** City planners are trying to decide among various parking-plan options ranging from more on-street spaces to multilevel facilities to spread-out small lots. Before making a decision, they wish to test the downtown merchants' claim that shoppers park for an average of only 47 minutes in the downtown area. The planners have decided to tabulate parking durations for 225 shoppers and to reject the merchants' claim if the sample mean exceeds 50 minutes. What is the probability of a Type II error if the true value is 48 minutes? If the true value is 51 minutes? Assume that the standard deviation in parking durations is 27 minutes.

***Answer:*** We have:

$$H_o: \; \mu = 47$$

$$H_a: \; \mu > 47$$

$$\sigma_{\bar{x}} = \sigma/\sqrt{n} = 27/\sqrt{225} = 1.8$$

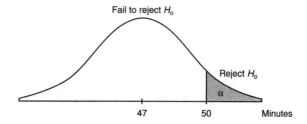

If the true mean parking duration is 48 minutes, the normal curve should be centered at 48. In this case, the $z$-score for 50 is $(50 - 48)/1.8 = 1.11$. Using Table A, we calculate the $\beta$-risk (probability of failure to reject $H_0$) to be $.5000 + .3665 = .8665$.

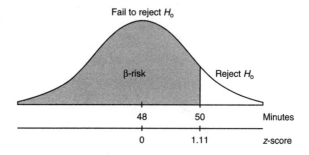

If the true mean parking duration is 51 minutes, then the normal curve should be centered at 51. The critical value is still 50, with a $z$-score now of $(50 - 51)/1.8 = -0.56$. We use Table A to find the $\beta$-risk to be $.5000 - .2123 = .2877$.

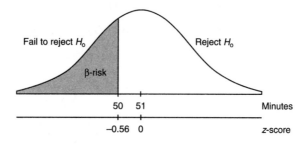

Note that the further the true value is (in the suspected direction) from the claimed mean, the smaller the probability is of failing to reject the false claim.

In many situations we start with a significance level, use this to calculate the critical value, and then find particular $\beta$-risks.

**Example 12.12** A geologist claims that a particular rock formation will yield a mean amount of 24 pounds of a chemical per ton of excavation. His company, fearful that the true amount will be less, plans to run a test on a sample of 50 tons.

    a. If the standard deviation between tons is 5.8 pounds, what is the critical value? Assume a 1% significance level.

***Answer:*** We have:

$$H_o: \ \mu = 24$$

$$H_a: \ \mu < 24$$

$$\alpha = 0.1$$

$$\sigma_{\bar{x}} \ = \ \frac{\sigma}{\sqrt{n}} \ = \ \frac{5.8}{\sqrt{50}} = 0.82$$

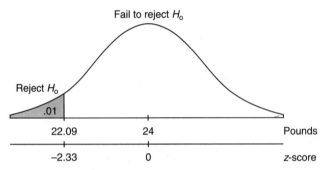

With $\alpha = .01$, the critical $z$-score is $-2.33$, and the critical poundage value is $24 - 2.33(0.82) = 22.09$. Thus, the decision is to reject $H_o$ if the sample mean is less than 22.09, and to fail to reject if it is more than 22.09.

    b. What is the probability of a Type II error if the true mean is 22? Is 20?

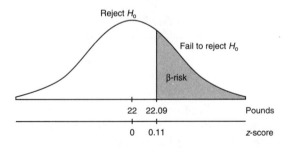

***Answer:*** If the true mean is 22 pounds per ton of rock, then the $z$-score for 22.09 is $(22.09 - 22)/0.82 = 0.11$, and so the $\beta$-risk is $.5000 - .0438 = .4562$.

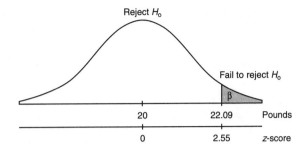

If the true mean is 20 pounds per ton of rock, then the $z$-score for 22.09 is $(22.09 - 20)/0.82 = 2.55$, and so the $\beta$-risk is $.5000 - .4946 = .0054$.

**Example 12.13** A medical research team claims that high vitamin C intake increases endurance. In particular, 1000 mg of vitamin C per day for a month should add an average of 4.3 minutes to the length of maximum physical effort that can be tolerated. Army training officers believe the claim is exaggerated and plan a test on 400 soldiers.

a. If the standard deviation of added minutes is 3.2, find the critical values for a significance level of 5% and then of 1%.

***Answer:*** We have:

$$H_o: \mu = 4.3$$

$$H_a: \mu < 4.3$$

$$\sigma_{\bar{x}} = 3.2/\sqrt{400} = 0.16$$

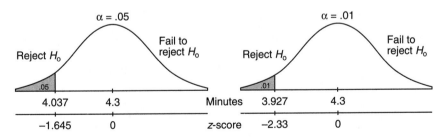

Corresponding to a $z$-score of $-1.645$ is $4.3 - 1.645(0.16) = 4.037$; corresponding to a $z$-score of $-2.33$ is $4.3 - 2.33(0.16) = 3.927$.

b. In each case, what is the probability of a Type II error if the true mean increase is only 4.2 minutes?

***Answer:***

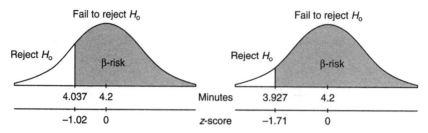

With a mean increase of 4.2, the $z$-score for 4.037 is $(4.037 - 4.2)/0.16 = -1.02$, and the $\beta$-risk is $.5000 + .3461 = .8461$. With a mean increase of 4.2, the $z$-score for 3.927 is $(3.927 - 4.2)/0.16 = -1.71$, and the $\beta$-risk is $.5000 + .4594 = .9564$.

Note that *decreasing* the $\alpha$-risk (from .05 to .01) led to an *increase* in the $\beta$-risk (from .8438 to .9564). This is always the case if the sample size is held constant.

**Example 12.14** A factory manager claims that the plant's smokestacks spew forth only 350 pounds of pollution per day. A government investigator suspects that the true value is higher and plans a hypothesis test with a critical value of 375 pounds. Suppose the standard deviation in daily pollution poundage is 150.

a. What is the $\alpha$-risk for a sample size of 100 days? Of 200 days?

***Answer:*** We have:

$$H_o: \ \mu = 350$$

$$H_a: \ \mu > 350$$

$$c = 375$$

$$n = 100 \qquad\qquad\qquad n = 200$$

$$\sigma_{\bar{x}} = \frac{150}{\sqrt{100}} = 15.0 \qquad\qquad \sigma_{\bar{x}} = \frac{150}{\sqrt{200}} = 10.6$$

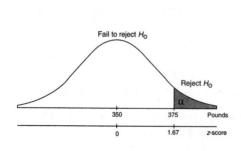

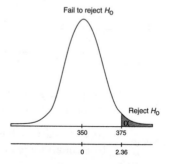

With $\sigma_{\bar{x}} = 15.0$, the $z$-score for 375 is $(375 - 350)/15.0 = 1.67$, and the corresponding $\alpha$-risk from Table A is $.5000 - .4525 = .0475$. With $\sigma_{\bar{x}} = 10.6$, the $z$-score for 375 is $(375 - 350)/10.6 = 2.36$, and the corresponding $\alpha$-risk is $.5000 - .4909 = .0091$.

b. What are the associated $\beta$-risks if the true mean is 385 pounds?

**Answer:**

We have:

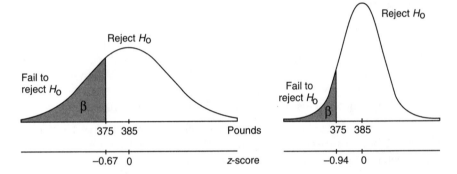

With $n = 100$ and a mean of 385, the z-score for 375 is $(375 - 385)/15.0 = -.67$ with a corresponding $\beta$-risk of $.5000 - .2486 = .2514$. With $n = 200$ and a mean of 385, the z-score for 375 is $(375 - 385)/10.6 = -.94$ with a corresponding $\beta$-risk of $.5000 - .3264 = .1736$.

Note that in this example *increasing* the sample size (from 100 to 200) resulted in *decreases* in both the $\alpha$-risk (from .0475 to .0091) and the $\beta$-risk (from .2514 to .1736). More generally, if $\alpha$ is held fixed and the sample size is increased, the $\beta$-risk will decrease. Furthermore, by increasing the sample size and adjusting the critical value, it is always possible to decrease both the $\alpha$-risk and the $\beta$-risk.

# EXERCISES

1. The label of a package of cords claims that the breaking strength of a cord is 3.5 pounds, but a hardware store owner believes the real value is less. She plans to test 36 such cords; if their mean breaking strength is less than 3.25 pounds, she will reject the claim on the label. If the standard deviation in breaking strengths of all such cords is 0.9 pounds, what is the probability of rejecting a true claim?

2. A union spokesman states that plumbers charge an average of $25 per

hour, but a consumers' bureau doubts that the pay rate is this low. A test is planned in which 40 plumbers will be asked to submit estimates on a job, and from these estimates a mean hourly rate will be calculated. Assuming a standard deviation among rates of $4.50, determine the critical value for a test at the 5% significance level.

3. A city spokesperson claims that the mean response time before a fire truck arrives at any fire is 12 minutes. A newspaper reporter suspects that the response time is actually longer and runs a test by examining the records of 64 fire emergency situations. What conclusion is reached if the sample mean is 13.1 minutes with a standard deviation of 6 minutes? Assume a 2% significance level.

4. A service station advertises that its mechanics will change a muffler in only 15 minutes. A consumers' group doubts this claim and runs a hypothesis test using 49 cars needing new mufflers. In this sample the mean changing time was 16.25 minutes with a standard deviation of 3.5 minutes.
   a. What is the $p$-value?
   b. What is the conclusion at a level of significance of 10%? 5%? 1%?

5. It is believed that using a new fertilizer will result in a yield of 1.6 tons per acre. A botanist carries out a two-tailed test on a field of 64 acres.
   a. Determine the $p$-value if the mean yield per acre in the sample is 1.72 tons with a standard deviation of 0.4.
   b. What is the conclusion at a level of significance of 10%? 5%? 1%?

6. A factory manager claims that the concentration of carbon monoxide in the air in the plant is only 3 parts per million. A government inspector believes the true figure is higher and gathers test data on 36 different days.
   a. If the resulting tabulations give $\Sigma x = 109.08$ and $\Sigma(x - \bar{x})^2 = .504$, what is the $p$-value?
   b. Can the manager's claim be rejected at a significance level of 1%? 2%? 5%?

7. The agent of a secretarial pool claims that their average speed is 85 words per minute, but this is disputed by the personnel manager of a company which is considering using the pool. A test is set up with a sample of 30 typists from the pool, and their words per minute figures tabulate as $\Sigma x = 2370$ and $\Sigma x^2 = 197,287$.
   a. What is the $p$-value?
   b. Should the agent's claim be rejected at a level of significance of 10%? 5%?

8. An author of a new book claims that anyone following his suggested diet program will lose an average of 2.8 pounds per week. A researcher believes that the true figure will be lower and plans a test involving 36 overweight people. She will reject the author's claim if the mean weight loss in the volunteer group is less than 2.5 pounds per week. What is the probability of a Type II error if the true value is 2.6? If the true value is

2.4? Assume that the standard deviation among individuals is 1.2 pounds per week.

9. A nursery owner predicts that a certain group of pine trees will grow an average of 25 centimeters per year, but a potential buyer is worried that the true rate will be lower. A test is set up with 100 trees.
   a. If the standard deviation between trees is 8 centimeters, what is the critical value? Assume a 1% significance level.
   b. What is the $\beta$-risk if the true mean growth rate is 24 cm? 22 cm? 20 cm?

10. A manufacturer claims that a motor will not draw more than 1.5 amperes under normal use conditions. A factory engineer suspects that this claim is low and plans to test 36 such motors. Suppose the standard deviation among motors is 0.9 amperes. If the true mean is 1.8 amperes, what is the probability of a Type II error associated with each of the following levels of significance: 10%, 5%, and 1%?

11. The following data from the National Bureau of Standards give daily radiation levels (in British thermal units) at 80 random North American locations: 1151, 236, 1107, 338, 848, 13, 142, 587, 555, 519, 506, 1106, 498, 946, 467, 651, 486, 599, 953, 426, 504, 332, 1248, 872, 66, 936, 713, 1037, 573, 848, 735, 524, 744, 890, 892, 526, 1149, 434, 899, 786, 1035, 590, 713, 704, 931, 912, 565, 119, 435, 1292, 1066, 590, 566, 540, 604, 938, 539, 1127, 566, 688, 1000, 633, 622, 1045, 984, 489, 603, 488, 592, 283, 252, 446, 502, 764, 1224, 451, 1172, 583, 632, and 488. The practicality of using roof-mounted solar collectors for home heating depends on the level of daily solar radiation available. Suppose that a company producing flat plate collectors needs an average daily radiation per square foot of horizontal surface of at least 750 Btu.
    a. What is the $p$-value?
    b. What conclusion should be reached at the 5% significance level? At the 1% significance level?

## Study Questions

1. What is involved in calculating the probability of a Type II error in a two-tailed test? For example, suppose that the average age of persons chosen for jury duty is believed to be 37. To test this, a court clerk keeps track of the ages of 100 jurors, notes a standard deviation of 12 years, and tests the mean at the 5% level of significance. What is the $\beta$-risk if the true mean age of jurors is 35?

2. We have seen that different possible values for the mean $\mu$ lead to different $\beta$-risks. If we plot $\mu$ along the $x$-axis and corresponding values of $\beta$ along the $y$-axis, the resulting graph is called the *operating characteristic curve*. Plot such a curve for Examples 12.11 and 12.12.

# 13
# DIFFERENCES IN POPULATION MEANS

Many real-life applications of statistics involve comparisons of *two* populations. For example, is the average weight of laboratory rabbits receiving a special diet greater than that of rabbits on a standard diet? Which of two accounting firms pays a higher mean starting salary? Is the life expectancy of a coal miner less than that of a school teacher? Some examples entail relationships between *three* or more populations; one technique for analyzing such cases will be considered in Chapter 15. In this chapter we concentrate on comparing two populations—more specifically, on comparing the *means* of the two populations.

First we consider how to compare the means of *samples*, one from each population. When is a difference between two such sample means significant? The answer is more apparent when we realize that what we are looking at is one difference from a set of differences. That is, there is the set of all possible differences obtained by subtracting sample means from one set from sample means of a second set. To judge the significance of one particular difference we must first determine how the differences vary among themselves. The necessary key, discussed in Chapter 2, is the fact that the variance of a set of differences is equal to the sum of the variances of the individual sets. Thus:

$$\sigma^2_{\bar{x}-\bar{y}} = \sigma^2_{\bar{x}} + \sigma^2_{\bar{y}}$$

Now, if

$$\sigma_{\bar{x}} = \frac{\sigma_1}{\sqrt{n_1}} \quad \text{and} \quad \sigma_{\bar{y}} = \frac{\sigma_2}{\sqrt{n_2}}$$

then

$$\sigma^2_{\bar{x}-\bar{y}} = \frac{\sigma_1^2}{n_1} + \frac{\sigma_2^2}{n_2} \quad \text{and} \quad \sigma_{\bar{x}-\bar{y}} = \sqrt{\frac{\sigma_1^2}{n_1} + \frac{\sigma_2^2}{n_2}}$$

# CONFIDENCE INTERVAL ESTIMATE FOR DIFFERENCE

In many situations it is clear that the mean of one population is higher than that of another, and we would like to estimate the difference. Using samples, we cannot find this difference *exactly*, but we will be able to say with a certain confidence that the difference lies in a certain *interval*. To find this *confidence interval estimate*, we follow the same procedure as set forth in Chapter 11 for a single mean, this time using $\mu_1 - \mu_2$, $\sqrt{(\sigma_1^2/n_1 + \sigma_2^2/n_2)}$, and $\bar{x}_1 - \bar{x}_2$ in place of $\mu$, $\sigma/\sqrt{n}$, and $\bar{x}$, respectively.

**Example 13.1** A 30-month study is conducted to determine the difference in the number of accidents per month between two departments in an assembly plant. Suppose the first department averages 12.3 accidents per month with a standard deviation of 3.5, while the second averages 7.6 accidents with a standard deviation of 3.4. Determine a 95% confidence interval estimate for the difference in the number of accidents per month.

***Answer:***

$$n_1 = 30 \qquad n_2 = 30$$

$$\bar{x}_1 = 12.3 \qquad \bar{x}_2 = 7.6$$

$$s_1 = 3.5 \qquad s_2 = 3.4$$

$$\sigma_{\bar{x}-\bar{y}} = \sqrt{\frac{\sigma_1^2}{n_1} + \frac{\sigma_2^2}{n_2}} \approx \sqrt{\frac{s_1^2}{n_1} + \frac{s_2^2}{n_2}} = \sqrt{\frac{(3.5)^2}{30} + \frac{(3.4)^2}{30}} = 0.89$$

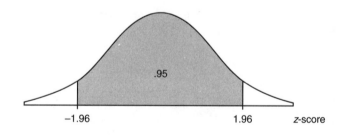

The observed difference is $12.3 - 7.6 = 4.7$, and the critical z-scores are $\pm 1.96$. Thus the confidence interval estimate is $4.7 \pm 1.96(0.89) = 4.7 \pm 1.74$. We are 95% confident that the first department has between 2.96 and 6.44 more accidents per month than does the second department.

**Example 13.2**  A survey is run to determine the difference in the cost of groceries in suburban stores versus inner-city stores. A preselected group of items is purchased in a sample of 45 suburban and 35 inner-city stores, and the following data are obtained:

| Suburban Stores | Inner-City Stores |
|---|---|
| $n_1 = 45$ | $n_2 = 35$ |
| $\bar{x}_1 = \$36.52$ | $\bar{x}_2 = \$39.40$ |
| $s_1 = \$1.10$ | $s_2 = \$1.23$ |

Find a 90% confidence interval estimate for the difference in the cost of groceries.

***Answer:***

$$\sigma_{\bar{x}-\bar{y}}M \approx \sqrt{\frac{(1.10)^2}{45} + \frac{(1.23)^2}{35}} = 0.265$$

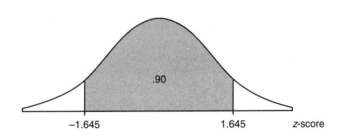

-1.645          1.645          z-score

The observed difference is $36.52 - 39.40 = -2.88$, the critical z-scores are $\pm 1.645$, and the confidence interval estimate is $-2.88 \pm 1.645(0.265) = -2.88 \pm 0.44$. Thus we are 90% certain that the selected group of items costs between $2.44 and $3.32 *less* in suburban stores than in inner-city stores.

**Example 13.3**  A trucking firm conducts a test to compare the life expectancies of two brands of tires. Shipments of 1000 and 1500 tires, respectively, are received and marked, and the truck mileages are noted when the tires are replaced. The resulting raw data are as follows:

| Brand $F$ | Brand $G$ |
|---|---|
| $n_1 = 1000$ | $n_2 = 1500$ |
| $\Sigma x = 22{,}350{,}000$ | $\Sigma x = 36{,}187{,}500$ |
| $\Sigma(x - \bar{x})^2 = 9{,}600{,}000{,}000$ | $\Sigma(x - \bar{x})^2 = 15{,}800{,}000{,}000$ |

Determine a 99% confidence interval estimate for the difference in life expectancies.

***Answer:*** Calculations of the means and standard deviations yield:

$$\bar{x}_1 = \frac{22{,}350{,}000}{1000} = 22{,}350 \qquad \bar{x}_2 = \frac{36{,}187{,}500}{1500} = 24{,}125$$

$$s_1 = \sqrt{\frac{9{,}600{,}000{,}000}{1000-1}} = 3100 \qquad s_2 = \sqrt{\frac{15{,}800{,}000{,}000}{1500-1}} = 3246.6$$

$$\sigma_{\bar{x}-\bar{y}} \approx \sqrt{\frac{(3100)^2}{1000} + \frac{(3246.6)^2}{1500}} = 129.0$$

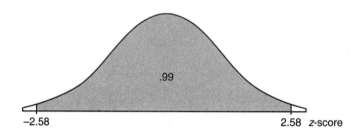

The observed difference is $22{,}350 - 24{,}125 = -1{,}775$, and the critical $z$-scores are $\pm 2.58$. Since $-1775 \pm 2.58(129) = -1775 \pm 333$, the trucking firm can be 99% sure that brand $F$ tires average between 1442 and 2108 miles less in life expectancy than brand $G$ tires.

**Example 13.4**  A hardware store owner wishes to determine the difference in drying times between two brands of paints. Suppose the standard deviation between cans in each population is 2.5 minutes. How large a sample (same number) of each must the store owner use if he wishes to be 98% sure of knowing the difference to within 1 minute?

***Answer:***

$$\sigma_{\bar{x}-\bar{y}} = \sqrt{\frac{(2.5)^2}{n} + \frac{(2.5)^2}{n}} = \frac{3.536}{\sqrt{n}}$$

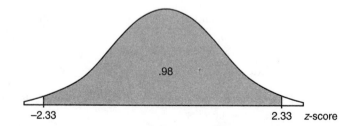

With a critical $z$-score of 2.33, we have $2.33(3.536/\sqrt{n}) \leq 1$, so $\sqrt{n} \geq 8.24$ and $n \geq 67.9$. Thus the owner should test samples of 68 paint patches from each brand.

# HYPOTHESIS TESTS: α-RISKS
# AND CRITICAL VALUES

In this situation the null hypothesis is usually that the means of the populations are the same, or, equivalently, that their difference is 0:

$$H_o: \mu_1 - \mu_2 = 0$$

The alternative hypothesis is then:

$$H_a: \mu_1 - \mu_2 < 0, \; H_a: \mu_1 - \mu_2 > 0, \quad \text{or } H_a: \mu_1 - \mu_2 \neq 0$$

The first two possibilities lead to one-sided (one-tailed) tests; the third possibility leads to two-sided (two-tailed) tests.

**Example 13.5**  A dentist believes that toothpaste brand $C$ is better than toothpaste brand $P$. She asks 30 of her patients to use each brand and records the numbers of cavities experienced over a 3-year period. She plans to reject any equality claim if the average number of cavities for brand $C$ users is at least 1 fewer than the average for brand $P$ users. If the standard deviations in cavities per person for each brand is 2.3, what is the probability of a Type I error, that is, of mistakenly rejecting a correct null hypothesis of equality?

**Answer:**

$$H_o: \mu_1 - \mu_2 = 0$$

$$H_a: \mu_1 - \mu_2 < 0$$

$$\sigma_{\bar{x}-\bar{y}} = \sqrt{\frac{(2.3)^2}{30} + \frac{(2.3)^2}{30}} = 0.594$$

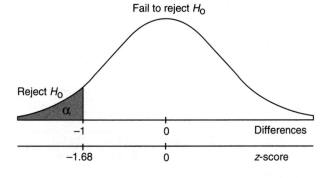

The $z$-score for $-1$ is $(-1-0)/0.594 = -1.68$. From Table A the resulting $\alpha$-risk is $.5000 - .4535 = .0465$.

**Example 13.6** A test is run to determine whether there is a difference in miles per gallon between two car models. Forty autos of the first model and 50 of the second are available for the test. What is the $\alpha$-risk if the cut-off difference scores are $\pm 0.5$? Assume standard deviations of 1.2 and 0.65 mpg, respectively, for the two models.

***Answer:***

$$H_o: \mu_1 - \mu_2 = 0$$

$$H_a: \mu_1 - \mu_2 \neq 0$$

$$\sigma_{\bar{x}-\bar{y}} \approx \sqrt{\frac{(1.2)^2}{40} + \frac{(0.65)^2}{50}} = 0.211$$

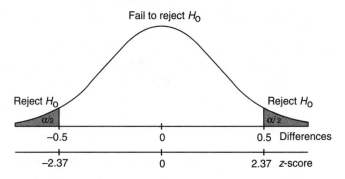

The $z$-score for 0.5 is $(0.5-0)/0.211 = 2.37$. From Table A, each tail has a probability of $.5000 - .4911 = .0089$. Adding the two tails gives an $\alpha$-risk of $.0089 + .0089 = .0178$.

**Example 13.7** A historian believes that the average height of soldiers in World War II was greater than that of soldiers in World War I. She examines the records of 100 men in each war and notes standard deviations of 2.5 and 2.3 inches in WWI and WWII, respectively. With an $\alpha$-risk of .05, what is the critical difference in heights?

***Answer:***

$$H_o: \mu_1 - \mu_2 = 0$$

$$H_a: \mu_1 - \mu_2 > 0$$

$$\alpha = .05$$

$$\sigma_{\bar{x}=\bar{y}} \approx \sqrt{\frac{(2.5)^2}{100} + \frac{(2.3)^2}{100}} = 0.340$$

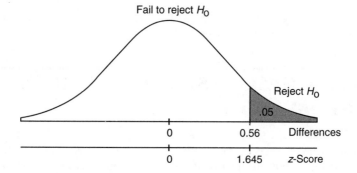

The critical $z$-score is 1.645, and the critical difference is $1.645(0.340) = 0.56$ inches.

**Example 13.8**    An educator plans to evaluate two different teaching methods by comparing the average test scores of two classes of students, each taught by a different one of the techniques under study. Suppose the two classes have 35 and 38 students respectively, with standard deviations in test scores of 24.5 and 21.2 respectively. What are the critical differences if the test is to be run with an $\alpha$-risk of .01?

***Answer:***

$$H_o : \mu_1 - \mu_2 = 0$$

$$H_a : \mu_1 - \mu_2 \neq 0$$

$$\alpha = .01$$

$$\sigma_{\bar{x}-\bar{y}} \approx \sqrt{\frac{(24.5)^2}{35} + \frac{(21.2)^2}{38}} = 5.38$$

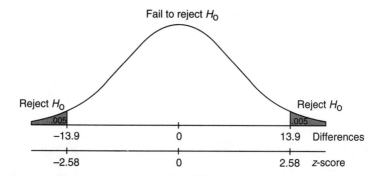

The critical values are $\pm 2.58(5.38) = \pm 13.9$. Thus, if the average test scores show a difference of more than 13.9, the educator will claim he has sufficient evidence that the two teaching methods yield different results.

# HYPOTHESIS TESTS: CONCLUSIONS AND p-VALUES

The fact that a sample mean from one population is greater than a sample mean from a second population does not automatically justify a similar conclusion about the means of the populations themselves. Two points need to be stressed. First, the means of samples from the same population vary from each other. Second, what we are really comparing are confidence interval estimates, not just single points.

**Example 13.9**   A sales representative believes that his company's computer has more average downtime per week than does a similar computer sold by a competitor. Before bringing his concern to his director, the sales representative gathers data and runs a hypothesis test. He determines that in a sample of 40 week-long periods in different firms using his company's product, the average downtime was 125 minutes with a standard deviation of 37 minutes. However, 35 week-long periods involving the competitor's computer yield an average downtime of only 115 minutes with a standard deviation of 43 minutes. What conclusion should the sales representative draw, assuming a 10% significance level?

**Answer:**

$$H_o: \mu_1 - \mu_2 = 0$$

$$H_a: \mu_1 - \mu_2 > 0$$

$$\alpha = .10$$

$$\sigma_{\bar{x}-\bar{y}} \approx \sqrt{\frac{(37)^2}{40} + \frac{(43)^2}{35}} = 9.33$$

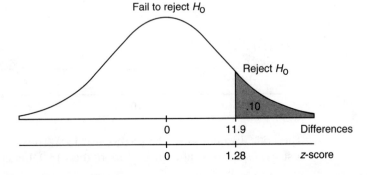

With $\alpha = .10$ the critical $z$-score is 1.28, so the critical difference is $0 + 1.28(9.33) = 11.9$ minutes. Since the observed difference is $125 - 115 = 10$, and $10 < 11.9$, the sales representative does *not* have sufficient evidence that his company's computer has more down time.

The strength of the observed difference in downtimes can be measured by finding the $p$-value. The z-score of the difference in sample means is $10/9.33 = 1.07$. Using Table A, we find the $p$-value to be $.5000 - .3577 = .1423$. Thus, while the observed difference is not significant at the 10% significance level, it would be significant at, for example, the 15% level.

**Example 13.10** A store manager wishes to determine whether there is a significant difference between two trucking firms with regard to the handling of egg cartons. In a sample of 200 cartons on one firm's truck, there was an average of 0.7 broken eggs per carton with a standard deviation of 0.31, while a sample of 300 cartons on the second firm's truck showed an average of 0.775 broken eggs per carton with a standard deviation of 0.42. Is the difference between the averages significant at a significance level of 5%? At a level of 1%?

***Answer:***

$$H_o: \mu_1 - \mu_2 = 0$$

$$H_a: \mu_1 - \mu_2 \neq 0$$

$$\alpha = .05$$

$$\sigma_{\bar{x}-\bar{y}} \approx \sqrt{\frac{(0.31)^2}{200} + \frac{(0.42)^2}{300}} = 0.0327$$

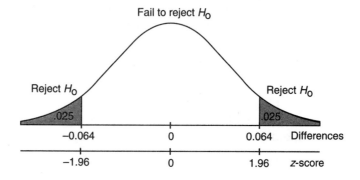

This is a two-tailed test, so, for a total $\alpha = .05$, we place a probability of $\frac{1}{2}(.05) = .025$ on each side. The critical $z$-scores are ±1.96, which translate into critical differences of 0 ± 1.96(0.0327) = ±0.064. Since the observed difference is 0.7 − 0.775 = −0.075 and −0.075 < −0.064, the difference in the average numbers of broken eggs per carton is significant at the 5% significance level.

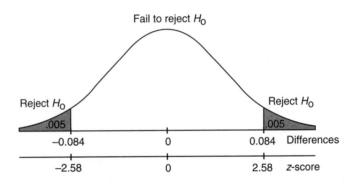

With α = .01, the critical z-scores are ±2.58, and so the critical differences are 0 ± 2.58(0.0327) = ±0.084. Since −0.075 is between −0.084 and 0.084, the observed difference is *not* significant at the 1% significance level.

A more specific measure of the strength of the observed difference in broken eggs per carton is found by calculating the *p*-value. The observed difference of −0.075 has a z-score of −0.075/0.0327 = −2.29. Using Table A, we find that this gives a probability of .5000 − .4890 = .0110. The test is two-sided, so the *p*-value is 2(.0110) = .0220.

**Example 13.11** A medical researcher believes that taking 1000 mg vitamin C per day will result in fewer colds than will a daily intake of 500 mg. In a group of 50 volunteers taking 1000 mg/day the numbers of colds per individual during a winter season are tabulated leading to

$$\Sigma x = 90 \text{ and } \Sigma(x - \bar{x})^2 = 73.5.$$

Similar tabulations from a group of 60 volunteers taking 500 mg/day give

$$\Sigma y = 144 \text{ and } \Sigma(y - \bar{y})^2 = 94.4.$$

a. If the experiment is run at the 2% significance level, what should be the conclusion?

***Answer:***

$$H_o: \mu_1 - \mu_2 = 0$$
$$H_a: \mu_1 - \mu_2 < 0$$
$$\alpha = .02$$

$$\bar{x} = \frac{\Sigma x}{50} = \frac{90}{50} = 1.8 \qquad\qquad \bar{y} = \frac{\Sigma y}{60} = \frac{144}{60} = 2.4$$

$$s_1^2 = \frac{\Sigma(x - \bar{x})^2}{50 - 1} = \frac{73.5}{49} = 1.5 \quad s_2^2 = \frac{\Sigma(y - \bar{y})^2}{60 - 1} = \frac{94.4}{59} = 1.6$$

$$\sigma_{\bar{x}-\bar{y}} \approx \sqrt{\frac{s_1^2}{50} + \frac{s_2^2}{60}} = \sqrt{\frac{1.5}{50} + \frac{1.6}{60}} = 0.238$$

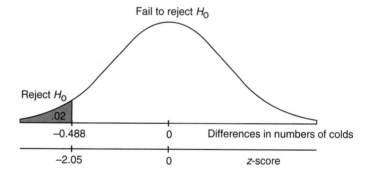

Fail to reject $H_0$

Reject $H_0$

.02

−0.488                    0         Differences in numbers of colds

−2.05                     0         z-score

For $\alpha = .02$ the critical $z$-score is $-2.05$, so the critical difference is $0 - 2.05(0.238) = -0.488$. The observed difference is $1.8 - 2.4 = -0.6$. Since $-0.6 < -0.488$, the researcher will conclude that fewer colds result when people take 1000 mg vitamin C per day rather than only 500 mg/day.

b. What is the *p*-value?

***Answer:*** The observed difference of $-0.6$ has a $z$-score of $-0.6/0.238 = -2.52$. Using Table A, we find a *p*-value of $.5000 - .4941 = .0059$.

**Example 13.12** A realtor believes that the average rent in one location exceeds the average rent in a second location by more than the published figure of $125. She runs a hypothesis test on 30 rental units in each location and obtains averages of $480 and $330, and standard deviations of $95 and $80, respectively, for the two locations. What should the realtor conclude at the 5% significance level?

***Answer:*** In this problem the null hypothesis is not that the difference is 0.

$$H_o: \mu_1 - \mu_2 = 125$$

$$H_a: \mu_1 - \mu_2 > 125$$

$$\alpha = .05$$

$$\sigma_{\bar{x}-\bar{y}} \approx \sqrt{\frac{(95)^2}{30} + \frac{(80)^2}{30}} = 22.68$$

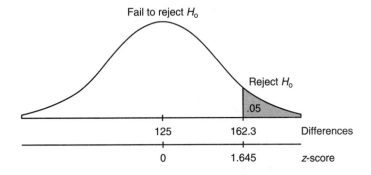

With $\alpha = .05$ the critical $z$-score is 1.645, so the critical difference is $125 + 1.645(22.68) = 162.3$. The observed difference is $480 - 330 = 150$. Since $150 < 162.3$, the realtor does *not* have sufficient evidence to claim that the rental difference is more than $125.

There was, however, some observed difference, and its strength can be measured by finding the *p*-value. The *z*-score of the observed difference of 150 is $(150 - 125)/22.68 = 1.10$. Using Table A, we find the *p*-value to be $.5000 - .3643 = .1357$.

# EXERCISES

1. In a study (*Journal of the American Medical Association,* June 13, 1990, page 3040) aimed at reducing developmental problems in low-birth-weight (under 2500 grams) babies, 347 were given a special educational curriculum while 561 did not receive any special help. After 3 years the children who had received the special curriculum showed a mean IQ of 93.5 with a standard deviation of 19.1; the other children had a mean IQ of 84.5 with a standard deviation of 19.9. Find a 95% confidence interval estimate for the difference in mean IQ's of low-weight babies who receive special intervention and those who do not.

2. Does socioeconomic status relate to age at time of HIV infection? For 274 high-income HIV-positive individuals the average age of infection was 33.0

years with a standard deviation of 6.3, while for 90 low-income individuals the average age was 28.6 years with a standard deviation of 6.3 (*The Lancet*, October 22, 1994, page 1121). Find a 90% confidence interval estimate for the difference in ages of high and low income people at the time of HIV infection.

3. An admissions director conducts a study to compare the scores of male and female students on the verbal section of a national standardized exam. Test results of 400 women and 500 men are noted to be as follows.

   Women: $n = 400$, $\Sigma x = 208,000$, $\Sigma(x - \bar{x})^2 = 760,000$
   Men: $n = 500$, $\Sigma x = 255,000$, $\Sigma(x - \bar{x})^2 = 1,050,000$

   Determine a 98% confidence interval estimate for the difference in mean verbal test results.

4. An engineer wishes to determine the difference in life expectancies between two brands of batteries. Suppose the standard deviation of each is 4.5 hours. How large a sample (same number) of each battery should be taken if the engineer wishes to be 90% certain of knowing the difference in life expectancy to within 1 hour?

5. A recent study of health service costs of coronary angioplasty (PTCA) versus coronary artery bypass surgery (CABG) at a London hospital (*The Lancet*, October 1, 1994, page 929) showed an average cost of 6176 pounds with a standard deviation of 329 for 231 angioplasties, and an average cost of 8164 pounds with a standard deviation of 264 for 221 bypass surgeries. Is this sufficient evidence at the 1% significance level to say that the average cost of angioplasties is less than the average cost of bypass surgeries?

6. Fifty-three children treated for lead poisoning had their IQ's tested before treatment and 6 months later (*Journal of the American Medical Association*, April 7, 1993, page 1644). Before treatment these children with unhealthy amounts of lead in their blood showed an average IQ of 83.5 with a standard deviation of 10.2; 6 months later their average IQ was 88.1 with a standard deviation of 11.2. Is this enough evidence to support the claim that average IQ improves after treatment for lead poisoning? Test with $\alpha = .05$.

7. An educator believes that professors at liberal arts colleges give higher grades than their colleagues in professional schools. She obtains a sample of 500 grades given at a liberal arts college and tabulates $\Sigma x = 1300$ and $\Sigma x^2 = 3460$, and a sample of 350 grades given at a professional school and tabulates $\Sigma x = 896$ and $\Sigma x^2 = 2360$. What conclusion should she draw at the 10% significance level?

8. A study reported in the *New York Times* (February 16, 1993, page C3) of roughly 75 left-handed and 925 right-handed people found the average

age of the left-handed people at death to be 66 years while the average age of the right-handed people at death was 75 years. What additional item of information is necessary to determine whether this is sufficient evidence at a 5% significance level to support the theory that right-handed people live longer than left-handed people? (This study was heavily criticized because it was based on recollections of friends and families and did not represent a good cross-section of the population.)

## Study Questions

1. (*A paired difference test*) The analysis and procedure developed in this chapter require that the two samples being compared be *independent* of each other. However, many experiments and tests involve comparing two populations for which the data naturally occur in pairs. In this case, the proper procedure is to run a one-sample test on the single variable consisting of the differences from the paired data.

   For example, suppose an efficiency expert wishes to analyze the difference in productivity between workers exposed to two different lighting arrangements. She chooses a procedure whereby the same 30 employees work under each arrangement for 1 week in turn. Each employee's output under the two lighting arrangements, LA1 and LA2, are summarized as follows.

| Emp | 1 | 2 | 3 | 4 | 5 | 6 | 7 | 8 | 9 | 10 | 11 | 12 | 13 | 14 | 15 |
|-----|----|----|----|----|----|----|----|----|----|----|----|----|----|----|----|
| LA1 | 15 | 23 | 18 | 19 | 27 | 13 | 22 | 20 | 20 | 19 | 25 | 26 | 15 | 19 | 19 |
| LA2 | 12 | 22 | 18 | 20 | 24 | 12 | 20 | 20 | 19 | 17 | 26 | 20 | 14 | 17 | 15 |

| | 16 | 17 | 18 | 19 | 20 | 21 | 22 | 23 | 24 | 25 | 26 | 27 | 28 | 29 | 30 |
|--|----|----|----|----|----|----|----|----|----|----|----|----|----|----|----|
| | 21 | 20 | 30 | 24 | 16 | 15 | 20 | 19 | 20 | 17 | 26 | 12 | 10 | 23 | 26 |
| | 16 | 22 | 26 | 24 | 19 | 19 | 18 | 19 | 15 | 20 | 22 | 14 | 13 | 20 | 25 |

   a. Form the set of thirty differences (be careful with signs).
   b. Run a single sample hypothesis test, $H_o$: $\mu = 0$, $H_a$: $\mu \neq 0$, $\alpha = .05$ on this set to determine whether the differences in lighting arrangements are significant.
   c. Find the *p*-value (you should get .0324).

2. a. Using the data given in Study Question 1, find the mean and standard deviation of each of the two 30-element sets corresponding to LA1 and LA2 separately.
   b. Run a two-sample hypothesis test, $H_o$: $\mu_1 - \mu_2 = 0$, $H_a$: $\mu_1 - \mu_2 \neq 0$, $\alpha = .05$.
   c. Find the *p*-value (you should get .3524).

   This seems to indicate that the observed difference is not significant in contrast with the analysis and conclusion in Study Question 1. The reason for this apparent inconsistency is that the test used in Study Question 2 is not the proper one. This two-sample test should be used only when the two sets are independent. In this case, there is a clear relationship between the data, in pairs, and this relationship is completely lost in the procedure for the two-sample test.

# 14
# SMALL SAMPLES

Statisticians and other researchers would save numerous man-hours if small samples could be used instead of larger ones. In many situations, such as estimating the completion time for a certain model of nuclear reactor or determining the damage sustained by luxury cars from front-end collisions, it is prohibitively difficult or expensive to gather enough data for a large sample. In other situations, such as deciding whether to add or drop a TV series after only a few shows have been aired, there is simply not sufficient time to gather large quantitiess of data.

## STUDENT *t*-DISTRIBUTION

If we have a normal population, then, for $n$ small, the set of means of all samples of size $n$ is still normally distributed, but the set of variances is not. The modified sample variances, $\Sigma(x - \bar{x})^2/(n - 1)$, do have a mean equal to the population variance, $\sigma^2$, but their distribution is skewed with more low values than high ones. Thus $(\bar{x} - \mu)/(s/\sqrt{n})$ has more high values (both plus and minus) and so does not follow a normal distribution. We can no longer say, for example, that 95% of the sample means are within 1.96 $(s/\sqrt{n})$ of the population mean. More than 5% of the sample means will lie outside these critical values.

If, however, the original population is normally distributed, there is a distribution that can be used when working with the $s/\sqrt{n}$ ratios. This *Student t-distribution*, was introduced in 1908 by W. S. Gosset, a British mathematician employed by the Guiness Breweries. (If we are working with small samples from a population that is *not* nearly normal, we must use very different "nonparametric" techniques not discussed in this book. If $n$ is large, we use the normal distribution tables.)

Thus, for a small sample from a normally distributed population, we work with the variable

$$t = \frac{\bar{x} - \mu}{s/\sqrt{n}}$$

with a resulting *t*-distribution that is bell-shaped and symmetrical, but is lower at the mean, higher at the tails, and so more spread out than the normal distribution.

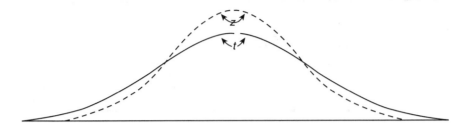

Like the binomial distribution, the *t*-distribution is different for different values of *n*. In the tables these distinct *t*-distributions are associated with the values for degrees of freedom (*df*). For this discussion the *df* value is equal to the sample size minus 1. The smaller the *df* value, the larger is the dispersion in the distribution. The larger the *df* value, that is, the larger the sample size, the closer the distribution comes to the normal distribution.

Since there is a separate *t*-distribution for each *df* value, fairly complete tables would involve many pages; therefore, in Table B of the Appendix we give areas and *t*-values for only the more commonly used percentages or probabilities. The last row of Table B is the normal distribution, which is a special case of the *t*-distribution taken when *n* is infinite. For practical purposes, the two distributions are very close for any $n \geq 30$. However, when more accuracy is required, the Student *t*-distribution can be used for much larger values of *n*.

Note that, whereas Table A gives areas under the normal curve from the mean to positive *z*-values, Table B gives areas to the right of given positive *t*-values. For example, suppose the sample size is 20, so $df = 20 - 1 = 19$. Then a probability of .05 in the *tail* corresponds with a *t*-value of 1.729, while .01 in the tail corresponds to $t = 2.539$.

Examples 14.1–14.7 involve small samples used for estimating population means, hypothesis testing on population means, and comparing two population means. *In all of these examples, it is assumed that the original populations are normally distributed.* Without this stipulation, we could not use the *t*-distribution, as explained above.

# SMALL SAMPLE ESTIMATES OF A POPULATION MEAN

**Example 14.1**   When ten cars of a new model were tested for gas mileage, the results showed a mean of 27.2 mpg with a standard deviation of 1.8 mpg. What is a 95% confidence interval estimate for the gas mileage achieved by this model?

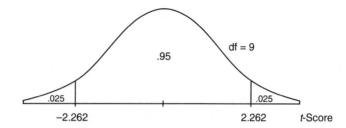

**Answer:** Because the sample size is less than 30, we must use the *t*-distribution. The standard deviation of the sample means is

$$\sigma_{\bar{x}} = \frac{1.8}{\sqrt{10}} = 0.569.$$

With $10 - 1 = 9$ degrees of freedom, and 2.5% in each tail, the appropriate *t*-scores are ±2.262. Thus we can be 95% certain that the gas mileage of the new model is in the range $27.2 \pm 2.262(0.569) = 27.2 \pm 1.3$, or between 25.9 and 28.5 miles per gallon.

**Example 14.2**  When 25 jars of peanut butter labeled as "18-ounce" are weighed, the totals are as follows: $\Sigma x = 448.5$ and $\Sigma(x - \bar{x})^2 = 0.41$. What is the 99% confidence interval estimate for the mean weight?

**Answer:** We first calculate the sample mean and standard deviation:

$$\bar{x} = \frac{\Sigma x}{n} = \frac{448.5}{25} = 17.94$$

$$s = \sqrt{\frac{\Sigma(x - \bar{x})^2}{n - 1}} = \sqrt{\frac{0.41}{24}} = 0.13$$

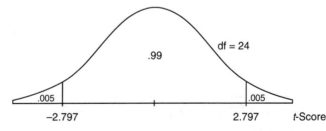

The standard deviation of the sample means is estimated to be

$$\sigma_{\bar{x}} \approx \frac{s}{\sqrt{n}} = \frac{0.13}{\sqrt{25}} = 0.026.$$

With $25 - 1 = 24$ degrees of freedom, and 0.5% in each tail, the appropriate *t*-scores are ±2.797. Thus a 99% confidence interval estimate for the mean weight is $17.94 \pm 2.797(0.026) = 17.94 \pm 0.073$ ounces.

**Example 14.3** A new process for producing synthetic gems yielded six stones weighing 0.43, 0.52, 0.46, 0.49, 0.60, and 0.56 carats, respectively, in its first run. Find a 90% confidence interval estimate for the mean carat weight from this process.

***Answer:***

$$\bar{x} = \frac{\Sigma x}{n} = \frac{0.43 + 0.52 + 0.46 + 0.49 + 0.60 + 0.56}{6}$$

$$= \frac{3.06}{6} = 0.51$$

$$s = \sqrt{\frac{\Sigma(x - \bar{x})^2}{n - 1}}$$

$$= \sqrt{\frac{(0.08)^2 + (0.01)^2 + (0.05)^2 + (0.02)^2 + (0.09)^2 + (0.05)^2}{5}}$$

$$= 0.0632$$

$$\sigma_{\bar{x}} \approx \frac{s}{\sqrt{n}} = \frac{0.0632}{\sqrt{6}} = 0.0258$$

With df $= 6 - 1 = 5$, and 5% in each tail, the $t$-scores are $\pm 2.015$. Thus we can be 90% sure that the new process will yield stones weighing $0.51 \pm 2.015(0.0258) = 0.51 \pm 0.052$, or between 0.458 and 0.562 carats.

# SMALL SAMPLE HYPOTHESIS TESTS OF A SAMPLE MEAN

**Example 14.4** A cigarette industry spokesperson remarks that current levels of tar are no more than 5 milligrams per cigarette. A reporter does a quick check on 15 cigarettes representing a cross section of the market.

a. What conclusion is reached if the sample mean is 5.63 milli-grams of tar with a standard deviation of 1.61? Assume a 10% significance level.

***Answer:***

$$H_o: \mu = 5$$

$$H_a: \mu > 5$$

$$\alpha = .10$$

$$\sigma_{\bar{x}} \approx \frac{s}{\sqrt{n}} = \frac{1.61}{\sqrt{15}} = 0.42$$

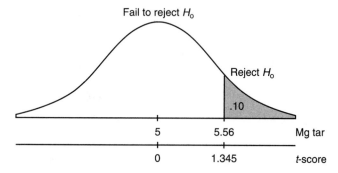

With df = 15−1 = 14, and $\alpha$ =.10, the critical *t*-score is 1.345. The critical number of milligrams of tar is 5 + 1.345(0.42) = 5.56. Since 5.63 > 5.56, the industry spokesperson's remarks should be rejected at the 10% significance level.

b. What is the conclusion at the 5% significance level?

***Answer:*** In this case, the critical *t*-score is 1.761; the critical tar level is 5 + 1.761(0.42) = 5.74 milligrams; and, since 5.63 < 5.74, the remarks cannot be rejected at the 5% significance level.

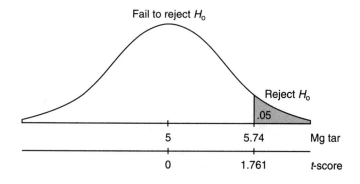

**Example 14.5** A local chamber of commerce claims that the mean sale price for homes in the city is $90,000. A real estate salesperson notes eight recent sales of $75,000, $102,000, $80,000, $85,000, $79,000, $95,000, $98,000, and $62,000. Should the chamber of commerce's claim be rejected at a significance level of 5%? At a level of 10%?

***Answer:***

$$H_o: \mu = 90,000$$

$$H_a: \mu \neq 90,000$$

$$\bar{x} = \frac{\Sigma x}{n} = \frac{676,000}{8} = 84,500$$

$$s = \sqrt{\frac{\Sigma(x - \bar{x})^2}{n - 1}} = \sqrt{\frac{1,246,000,000}{7}} = 13,341.7$$

$$\sigma_{\bar{x}} \approx \frac{s}{\sqrt{n}} = \frac{13,341.7}{\sqrt{8}} = 4717$$

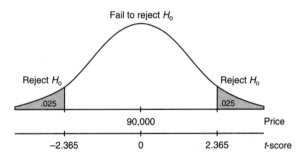

Fail to reject $H_o$

Reject $H_o$          Reject $H_o$

.025        .025

90,000     Price

−2.365     0     2.365     *t*-score

For $\alpha = .05$, we place a probability of .025 in each tail, and note that df $= 8 - 1 = 7$ gives critical *t*-scores of $\pm 2.365$. Critical sales prices are $90,000 \pm 2.365(4717) = 90,000 \pm 11,156$, or $78,844 and $101,156. Since the observed mean of $84,500 is in this range, there is *not* sufficient evidence to reject the claim at the 5% significance level.

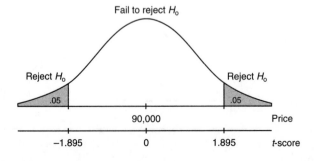

Fail to reject $H_o$

Reject $H_o$          Reject $H_o$

.05        .05

90,000     Price

−1.895     0     1.895     *t*-score

For $\alpha = .10$, we place a probability of .05 in each tail, and note critical $t$-scores of $\pm 1.895$. Critical sales prices are $90,000 \pm 1.895(4717) = 90,000 \pm 8939$, or \$81,061 and \$98,939. The observed mean of \$84,500 is also in this range, so even at the 10% significance level the claim cannot be rejected.

# SMALL SAMPLE DIFFERENCES IN POPULATION MEANS

In this case, we must assume not only that both original populations are normally distributed, but also that their variances are equal. Then for $\sigma_{\bar{x}-\bar{y}}$ (see Chapter 13) we get

$$\sigma_{\bar{x}-\bar{y}} = \sqrt{\frac{\sigma_1^2}{n_1} + \frac{\sigma_2^2}{n_2}} = \sqrt{\frac{\sigma^2}{n_1} + \frac{\sigma^2}{n_2}} = \sigma\sqrt{\frac{1}{n_1} + \frac{1}{n_2}}$$

But what should we use for $\sigma$? The idea, motivated by

$$s^2 = \frac{\Sigma(x - \bar{x})^2}{n - 1} \qquad \text{or} \qquad \Sigma(x - \bar{x})^2 = (n - 1)s^2, \qquad \text{is to use}$$

$$\frac{\Sigma(x - \bar{x}_1)^2 + \Sigma(x - \bar{x}_2)^2}{(n_1 - 1) + (n_2 - 1)}$$

for $s^2$, where the first sum is over the first sample and the second sum is over the second sample. This quotient can be rewritten as

$$\frac{(n_1 - 1)s_1^2 + (n_2 - 1)s_2^2}{n_1 + n_2 - 2}$$

and finally we have the complex-looking equation

$$\sigma_{\bar{x}-\bar{y}} \approx \sqrt{\frac{(n_1 - 1)s_1^2 + (n_2 - 1)s_2^2}{n_1 + n_2 - 2}} \sqrt{\frac{1}{n_1} + \frac{1}{n_2}}$$

The degrees of freedom in such a situation are

$$df = (n_1 - 1) + (n_2 - 1) = n_1 + n_2 - 2.$$

**Example 14.6**   Two varieties of corn are being compared as to difference in maturation time. Ten plots of the first variety reach maturity in an average of 95 days with a standard deviation of 5.3 days, while eight plots of the second variety reach maturity in an average of 74 days with a standard deviation of 4.8 days. Determine a 95% confidence interval estimate for the difference in maturation time.

***Answer:***

$$n_1 = 10 \qquad n_2 = 8$$

$$\bar{x}_1 = 95 \qquad \bar{x}_2 = 74$$

$$s_1 = 5.3 \qquad s_2 = 4.8$$

$$\sigma_{\bar{x}-\bar{y}} \approx \sqrt{\frac{(10-1)(5.3)^2 + (8-1)(4.8)^2}{10+8-2}} \sqrt{\frac{1}{10} + \frac{1}{8}} = 2.41$$

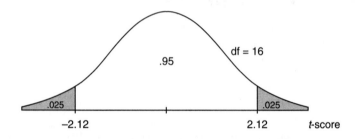

.95    df = 16

.025    .025

−2.12    2.12    *t*-score

With df $= 10 + 8 - 2 = 16$, and .025 in each tail, the critical *t*-scores are $\pm 2.12$. The observed difference was $95 - 74 = 21$, so the confidence interval estimate is $21 \pm 2.12(2.41) = 21 \pm 5.1$. Thus, we can be 95% certain that the first variety of corn will take between 15.9 and 26.1 more days to reach maturity than will the second variety.

**Example 14.7**   A city council member claims that male and female officers wait equal times for promotion in the police department. A women's spokesperson, however, believes women must wait longer than men. If five men waited 8, 7, 10, 5, and 7 years for promotion while four women waited 9, 5, 12, and 8 years, what conclusion should be drawn at the 10% significance level?

***Answer:***

$$H_o: \mu_1 - \mu_2 = 0$$

$$H_a: \mu_1 - \mu_2 < 0$$

$$\alpha = .10$$

$$n_1 = 5 \qquad \bar{x}_1 = \frac{8 + 7 + 10 + 5 + 7}{5} = 7.4$$

$$s_1 = \sqrt{\frac{(8-7.4)^2 + (7-7.4)^2 + (10-7.4)^2 + (5-7.4)^2 + (7-7.4)^2}{5-1}} = 1.82$$

$$n_2 = 4 \qquad \bar{x}_2 = \frac{9 + 5 + 12 + 8}{4} = 8.5$$

$$s_2 = \sqrt{\frac{(9 - 8.5)^2 + (5 - 8.5)^2 + (12 - 8.5)^2 + (8 - 8.5)^2}{4 - 1}} = 2.89$$

$$\sigma_{\bar{x}-\bar{y}} \approx \sqrt{\frac{(5 - 1)(1.82)^2 + (4 - 1)(2.89)^2}{5 + 4 - 2}} \sqrt{\frac{1}{5} + \frac{1}{4}} = 1.57$$

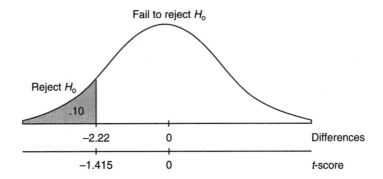

With df $= 5 + 4 - 2 = 7$, and .10 in the tail, the critical $t$-score is $-1.415$. The critical difference is $-1.415(1.57) = -2.22$. Since the observed difference is only $7.4 - 8.5 = -1.1$, there is *not* sufficient evidence to dispute the council member's claim.

# EXERCISES

In the following exercises, there is the underlying assumption that *all original populations have nearly normal distributions*. In Exercises 11–15, there is the additional assumption that the two original variances are equal.

1. In a sleep-laboratory experiment, 16 volunteers sleep an average of 7.4 hours with a standard deviation of 1.3 hours. What would be a 99% confidence interval estimate for the mean number of hours that people sleep at night?

2. A doctor notes the number of hours that 17 surgery patients spend in the recovery room after use of a new anesthetic. If $\Sigma x = 35.7$ and $\Sigma(x - \bar{x})^2 = 0.096$, what is an 80% confidence interval estimate for the mean number of hours spent in the recovery room by patients using the new anesthetic?

3. Acute renal graft rejection can happen years after the graft. In one study (*The Lancet*, December 24, 1994, page 1737) 21 patients showed such late-acute rejection when the ages of their grafts (in years) were 9, 2, 7, 1, 4, 7, 9, 6, 2, 3, 7, 6, 2, 3, 1, 2, 3, 1, 1, 2, and 7. Establish a 90% confidence

interval estimate for the ages of renal grafts that undergo late-acute rejection.

4. Nine subjects, 87 to 96 years old, were given 8 weeks of progressive-resistance weight training (*Journal of the American Medical Association*, June 13, 1990, page 3032). Strength before and after training for each individual was measured as maximum weight in (kilograms) lifted by left knee extension:

| Before: | 3 | 3.5 | 4 | 6 | 7 | 8 | 8.5 | 12.5 | 15 |
|---|---|---|---|---|---|---|---|---|---|
| After: | 7 | 17 | 19 | 12 | 19 | 22 | 28 | 20 | 28 |

Find 95% confidence intervals estimates for:
a. Before training strength
b. After training strength
c. Strength gain

5. An IRS representative claims that the average deduction for medical care is $1250. A taxpayer who believes that the real figure is lower samples 12 families and comes up with a mean of $934 and a standard deviation of $616. What conclusion should the taxpayer reach at a 5% significance level?

6. An auditor remarks that the accounts receivable for a company seem to average about $2000. A quick check of 20 accounts gives a mean of $2250 with a standard deviation of $600. Is this sufficient evidence to dispute the auditor's remark at a 5% significance level?

7. A magazine article states that teenagers watch 2 hours of television per day. A sociologist who believes the true figure is higher interviews 18 teenagers and calculates $\Sigma x = 50.4$ and $\Sigma x^2 = 163.8$. What is the conclusion at the 1% significance level?

8. PCB contamination of a river by a manufacturer is being measured by amounts of the pollutant found in fish. A company scientist claims that the fish contain only 5 parts per million, but an investigator believes the figure is higher. What is the conclusion if six fish are caught and show the following amounts of PCB (in parts per million): 6.8, 5.6, 5.2, 4.7, 6.3, and 5.4? Test at the 5% significance level.

9. The 26 contestants in the Pentathlon event of the 1992 Olympics ran the 200 meter dash in the following times (in seconds): 25.44, 24.39, 25.66, 23.93, 23.34, 25.01, 24.27, 24.54, 25.44, 24.86, 23.95, 23.31, 24.60, 23.12, 25.29, 24.40, 25.24, 24.43, 23.70, 25.20, 25.09, 24.48, 26.13, 25.28, 24.18, and 23.83 (*Journal of the American Statistical Association*, September 1994, page 1101). At the 1% significance level, test the null hypothesis that the mean time for Pentathlon contestants is 25 seconds against the alternative hypothesis that the mean time is less than 25 seconds.

10. The weight of an aspirin tablet is 300 mg according to the bottle label. Should an FDA investigator reject the label if she weighs seven tablets and obtains weights of 299, 300, 305, 302, 299, 301, and 303 milligrams? Use a 2.5% significance level for this two-tailed test.

11. Six capsules of drug *A* took an average of 75 seconds with a standard deviation of 1.4 seconds to dissolve, while the average time for six capsules of drug *B* was 71 seconds with a standard deviation of 1.7 seconds. Establish a 99% confidence interval estimate for the difference in dissolving time between the two brands.

12. The mean height of eight dwarf apple trees is 8.1 feet with a standard deviation of 2.3 feet; for five dwarf cherry trees the mean height is 9.8 feet with a standard deviation of 2.8 feet. What is the 95% confidence interval estimate of the difference in heights between these two fruit trees?

13. A hospital exercise-laboratory technician notes the resting pulse rates of five joggers to be 60, 58, 59, 61, and 67, while the resting pulse rates of seven nonexercisers are 83, 60, 75, 71, 91, 82, and 84. Establish a 90% confidence interval estimate for the difference in pulse rates between joggers and nonexercisers.

14. A researcher believes a new diet should improve weight gain in laboratory mice. If ten control mice on the old diet gain an average of 4 ounces with a standard deviation of 0.3 ounces, while the average gain for ten mice on the new diet is 4.8 ounces with a standard deviation of 0.2 ounces, is the researcher's claim justified at the 5% significance level?

15. An employer wishes to compare typing speeds of graduates from two different programs of study. Eight graduates of the first course type at 62, 85, 59, 64, 73, 70, 75, and 72 words per minute, while six graduates of the second course have typing speeds of 75, 64, 81, 55, 69, and 58 words per minute. Is the observed difference significant with $\alpha = .20$?

## Study Question

(*A paired difference test*) An accounting firm measured the blood pressures of ten of its CPAs before and then during the spring 1996 tax season. The systolic pressures for the ten individuals, designated as $A - J$, were as follows:

|          | A   | B   | C  | D   | E   | F   | G   | H   | I   | J  |
|----------|-----|-----|----|-----|-----|-----|-----|-----|-----|----|
| Before:  | 110 | 124 | 98 | 105 | 115 | 120 | 118 | 110 | 123 | 95 |
| During:  | 115 | 126 | 97 | 108 | 115 | 124 | 119 | 113 | 121 | 96 |

Is there sufficient evidence that blood pressures rise during tax season?
a. Assuming normal populations with equal variances, try to apply the two-population, small-sample hypothesis test developed in this chapter on $H_0$: $\mu_1 - \mu_2 = 0$, $H_a$: $\mu_1 - \mu_2 < 0$. Show that this test does *not* indicate that the observed difference is significant even at $\alpha = .10$.
b. Looking at the data, however, it is apparent that a change in blood

pressure occurred. The test used in part *a* does not apply the knowledge of what happened to each individual. In fact, since the before and during sets are not independent, but rather are related in pairs, that test is not the proper one to use. The appropriate test is a one-population, small-sample hypothesis test on the set of differences. Form the set of ten differences, being careful of sign. Then test the hypotheses $H_0$: $\mu = 0$, $H_a$: $\mu < 0$, and show that the sample mean is significant at $\alpha = .05$.

# 15
# COMPARING MORE THAN TWO MEANS

We now consider cases in which three or more populations are being compared. A sample is taken from each population, and the mean of each sample is calculated. When can we say that the observed differences in the sample means are significant? When should we say that there is insufficient evidence that the actual population means are different?

One method would be to use the $t$-distribution to compare the first sample mean with the second, the second with the third, and so on. It is not obvious, however, how to combine these results into a statement about $H_0$: $\mu_1 = \mu_2 = \mu_3 = \ldots$.

## VARIANCE BETWEEN AND WITHIN SAMPLES

Another approach is suggested by the following example. Consider these samples taken from three different populations:

I. {10.1, 10.8, 9.7, 9.3, 10.1} with mean 10
II. {14.2, 14.9, 15.3, 15.4, 15.2} with mean 15
III. {20.5, 19.8, 19.9, 20.1, 19.7} with mean 20

Intuitively, there seems to be sufficient evidence to conclude that the means of the three original populations are probably not identical. However, suppose that the samples had been as follows:

I. {2, 25, −14, −3, 40} with mean 10
II. {−4, 82, 0, −15, 12} with mean 15
III. {42, −30, 87, 0, 1} with mean 20

Here, there does not at all seem to be sufficient evidence that the means of the three original populations are different. Note that in the first example the differences *between* the sample means are large when compared to the variations *within* each sample, whereas in the second example the differences *between* the sample means are very small when compared to the variations *within* the samples.

Given that differences observed *between* sample means are more significant when they are large compared to differences *within* each sample, we consider a ratio of the form:

$$\frac{\text{differences between}}{\text{differences within}}$$

The larger such a ratio, the more evidence that the original population means are different. What should go into a ratio of this kind? The logical choice for measuring differences is *variance*. Therefore, we will define and calculate a "variance between" and a "variance within," and then form a ratio of these two variances. If this ratio, called the *F-ratio*, is greater than some critical *F*-value, we will conclude that there is sufficient evidence to reject $H_o$: $\mu_1 = \mu_2 = \mu_3 = \dots$.

**Example 15.1** Following are monthly sales figures (number of houses sold) for 15 real estate agents who work under three different types of contract:

| | |
|---|---|
| Straight salary: | 4, 7, 8, 5, 6 |
| Straight commission: | 7, 2, 4, 5, 7 |
| Salary plus commission: | 7, 9, 6, 6, 7 |

Calculation of means for the three groups yields

$$\bar{x}_1 = \frac{30}{5} = 6 \qquad \bar{x}_2 = \frac{25}{5} = 5 \qquad \bar{x}_3 = \frac{35}{5} = 7$$

Are these observed differences significant? Is there sufficient evidence to reject $H_o$: $\mu_1 = \mu_2 = \mu_3$?

***Answer:*** The variance *within* samples, $\sigma^2_{\text{within}}$ , is calculated by averaging the individual variances:

$$s_1^2 = \frac{(4-6)^2 + (7-6)^2 + (8-6)^2 + (5-6)^2 + (6-6)^2}{5-1}$$

$$= \frac{10}{4} = 2.5$$

$$s_2^2 = \frac{(7-5)^2 + (2-5)^2 + (4-5)^2 + (5-5)^2 + (7-5)^2}{5-1}$$

$$= \frac{18}{4} = 4.5$$

$$s_3^2 = \frac{(7-7)^2 + (9-7)^2 + (6-7)^2 + (6-7)^2 + (7-7)^2}{5-1}$$

$$= \frac{6}{4} = 1.5$$

$$\sigma^2_{\text{within}} = \frac{2.5 + 4.5 + 1.5}{3} = 2.83$$

The variance *between* samples, $\sigma^2_{\text{between}}$, is calculated by first determining $\sigma^2_{\bar{x}}$ (the variance of the sample means), and then using $\sigma^2_{\text{between}} = n\sigma^2_{\bar{x}}$ (from $\sigma_{\bar{x}} = \sigma/\sqrt{n}$):

$$\mu = \frac{\bar{x}_1 + \bar{x}_2 + \bar{x}_3}{3} = \frac{6 + 5 + 7}{3} = 6$$

$$\sigma^2_{\bar{x}} = \frac{(6-6)^2 + (5-6)^2 + (7-6)^2}{3-1} = \frac{2}{2} = 1$$

$$\sigma^2_{\text{between}} = n\sigma^2_{\bar{x}} = (5)(1) = 5$$

The *F-ratio* is defined to be

$$F = \frac{\sigma^2_{\text{between}}}{\sigma^2_{\text{within}}} = \frac{5}{2.83} = 1.76$$

Is this *F*-ratio large enough to enable us to reject $H_o$?

To answer this we must first find a critical *F*-value. Like the *t*-distribution, the *F*-distribution is actually a family of distributions. Where the *t*-distributions are associated with single degrees of freedom, however, the *F*-distributions are associated with *pairs* of degrees of freedom. There is a *df* value for the numerator, and a second *df* value for the denominator.

df of the numerator    = number of samples $-1 = k-1$
df of the denominator = (number of samples)(sample size $-1$)
                            = $k(n-1)$

Separate tables exist for each level of significance under consideration.

In this example, there are $3 - 1 = 2$ degrees of freedom for the numerator, and $3(5 - 1) = 12$ degrees of freedom for the denominator. Using, for example, $\alpha = .05$, we find a critical *F*-value of 3.88. Since $1.76 < 3.88$, there is not sufficient evidence to reject $H_o$.

# FORMAL PROCEDURE TO COMPARE SEVERAL SAMPLE MEANS

Just as in using the *t*-distribution to compare two sample means, we must be able to assume, when using the *F*-distribution, that all original populations are normal, and that their variances are equal. Then, if we have *k* populations and a sample of size *n* from each, we test the null hypothesis as follows.

$$H_o: \mu_1 = \mu_2 = \dots \mu_k$$

1. Calculate the sample means: $\bar{x}_1, \bar{x}_2, ..., \bar{x}_k$ where

$$\bar{x}_i = \frac{\Sigma x}{n}$$

2. Calculate the sample variances: $s_1^2, s_2^2, ..., s_k^2$ where

$$s_i^2 = \frac{\Sigma(x - \bar{x}_i)^2}{n - 1}$$

3. Calculate the overall mean:

$$\mu = \frac{\bar{x}_1 + \bar{x}_2 + ... + \bar{x}_k}{k}$$

4. Calculate the variance of the sample means:

$$\sigma_{\bar{x}}^2 = \frac{(\bar{x}_1 - \mu)^2 + (\bar{x}_2 - \mu)^2 + ... + (\bar{x}_k - \mu)^2}{k - 1}$$

5. Calculate the variance within samples:

$$\sigma_{within}^2 = \frac{s_1^2 + s_2^2 + ... + s_k^2}{k}$$

6. Calculate the variance between samples:

$$\sigma_{between}^2 = n\sigma_{\bar{x}}^2$$

7. Calculate the $F$-ratio: $\dfrac{\sigma_{between}^2}{\sigma_{between}^2}$

8. Find the critical $F$-value by using the appropriate $F$-table (depending on significance level), and looking under $df_{numerator} = k - 1$ and $df_{denominator} = k(n - 1)$.

9 If $F$-ratio $>$ $F$-critical, be aware that there is sufficient evidence to reject $H_o$. Otherwise, fail to reject $H_o$.

The above procedure, called *one-factor* or *one-way* or *one-dimensional analysis of variance*, was developed in the 1920s by Ronald Fisher, and the $F$s in $F$-score, $F$-ratio, and $F$-table are derived from his name. Similar, but more involved, methods are available when two factors have effects and are under consideration.

Here is a second example of one-way analysis of variance (abbreviated ANOVA).

**Example 15.2**   A consumer spokesperson claims that the price of a gallon of milk in a major city varies by location. She notes the prices in four stores chosen in each of five locations.

> Location 1: 1.98, 2.01, 2.02, 1.99
> Location 2: 2.04, 2.03, 2.04, 2.07
> Location 3: 2.06, 2.08, 2.08, 2.08

Location 4: 1.95, 1.97, 1.96, 1.98
Location 5: 2.00, 1.99, 2.02, 2.01

Is there sufficient evidence to reject the null hypothesis $H_o$: $\mu_1 = \mu_2 = \mu_3 = \mu_4 = \mu_5$ at the 1% significance level?

**Answer:** The sample means are as follows:

$$\bar{x}_1 = \frac{1.98 + 2.01 + 2.02 + 1.99}{4} = 2.000$$

$$\bar{x}_2 = \frac{2.04 + 2.03 + 2.04 + 2.07}{4} = 2.045$$

$$\bar{x}_3 = \frac{2.06 + 2.08 + 2.08 + 2.08}{4} = 2.075$$

$$\bar{x}_4 = \frac{1.95 + 1.97 + 1.96 + 1.98}{4} = 1.965$$

$$\bar{x}_5 = \frac{2.00 + 1.99 + 2.02 + 2.01}{4} = 2.005$$

The sample variances are as follows:

$$s_1^2 = \frac{(1.98 - 2)^2 + (2.01 - 2)^2 + (2.02 - 2)^2 + (1.99 - 2)^2}{4 - 1}$$
$$= \frac{0.001}{3} = 0.0003333$$

$$s_2^2 = \frac{(2.04 - 2.045)^2 + ... + (2.07 - 2.045)^2}{4 - 1}$$
$$= \frac{0.0009}{3} = 0.0003$$

$$s_3^2 = \frac{(2.06 - 2.075)^2 + ... + (2.08 - 2.075)^2}{4 - 1}$$
$$= \frac{0.0003}{3} = 0.0001$$

$$s_4^2 = \frac{(1.95 - 1.965)^2 + ... + (1.98 - 1.965)^2}{4 - 1}$$
$$= \frac{0.0005}{3} = 0.0001667$$

$$s_5^2 = \frac{(2.00 - 2.005)^2 + ... + (2.01 - 2.005)^2}{4 - 1}$$
$$= \frac{0.0005}{3} = 0.0001667$$

The overall mean is

$$\mu = \frac{2.000 + 2.045 + 2.075 + 1.965 + 2.005}{5} = 2.018$$

The variance of sample means is

$$\sigma_{\bar{x}}^2 = \frac{(2 - 2.018)^2 + (2.045 - 2.018)^2 + \ldots + (2.005 - 2.018)^2}{5 - 1}$$

$$= \frac{0.00728}{4} = 0.00182$$

The variance within samples is

$$\sigma_{within}^2 = \frac{0.0003333 + 0.0003 + 0.0001 + 0.0001667 + 0.0001667}{5}$$

$$= 0.0002133$$

The variance between samples is

$$\sigma_{between}^2 = n\sigma_{\bar{x}}^2 = 4(0.00182) = 0.00728$$

The $F$-ratio is

$$F = \frac{\sigma_{between}^2}{\sigma_{within}^2} = \frac{0.00728}{0.0002133} = 34.13$$

Since $df_{numerator} = k - 1 = 5 - 1 = 4$, $df_{denominator} = k(n - 1) = 5(4 - 1) = 15$, and $\alpha = .01$, the critical $F$-value is $F_{critical} = 4.90$. Then, since $34.13 > 4.90$, there is sufficient evidence to reject $H_o$. The consumer spokesperson is justified in claiming that the price of a gallon of milk varies by location.

# EXERCISES

1. Lime tree honey (L), fruit tree honey (F), and acacia honey (A) were tested for healing properties against each of four bacteria strains (*The Lancet*, March 20, 1993, page 757). Measurements of the effectiveness of the honeys on a 0 to 5 scale were as follows:

   L: 4, 4, 4, 3        F: 1, 2, 2, 0        A: 3, 4, 3, 1

   At a 5% significance level is there sufficient evidence that the three honeys do not have the same effectiveness as healing agents?

2. A double-blind test was run to judge the effectiveness of a new drug (cA2) in treating rheumatoid arthritis (*The Lancet*, October 22, 1994, page 1107). Of a group of 75 patients, 25 were given a placebo, 25 were given low doses of cA2, and 25 were given high doses of cA2. After 4 weeks of treatment, the patients' swollen joints were measured on a 1 to 60 scale with the following results:

   Placebo:            $\bar{x}_1 = 23.0$    with $s_1 = 11.2$
   Low dose of cA2:    $\bar{x}_2 = 12.9$    with $s_2 = 8.8$
   High dose of cA2:   $\bar{x}_3 = 8.6$    with $s_3 = 6.4$

At a 1% significance level is there sufficient evidence that the three treatments produce different results?

3. An executive authorizes a study to examine the length of time between order and service of personal pan pizzas in different franchise units. In each of five restaurants, 15 lengths of times (in minutes) are observed.

| Franchise 1 | Franchise 2 | Franchise 3 | Franchise 4 | Franchise 5 |
|---|---|---|---|---|
| 3 | 4 | 4 | 3 | 4 |
| 2.5 | 3.5 | 3.5 | 3.5 | 3.5 |
| 3 | 5 | 4 | 5 | 5 |
| 4 | 4.5 | 4 | 5 | 5 |
| 3.5 | 4 | 4 | 3.5 | 4 |
| 6 | 4 | 3 | 3.5 | 6.5 |
| 3.5 | 3.5 | 3.5 | 4 | 4 |
| 4 | 4.5 | 5 | 3 | 4.5 |
| 2.5 | 5 | 4.5 | 4.5 | 5 |
| 5 | 5.5 | 6 | 3.5 | 3.5 |
| 4 | 4 | 3.5 | 3 | 6 |
| 4 | 3.5 | 3 | 3 | 4.5 |
| 3.5 | 4.5 | 4 | 4.5 | 4 |
| 4 | 4 | 3.5 | 4 | 4 |
| 3.5 | 4.5 | 4 | 3.5 | 3.5 |

Is the executive justified in concluding that different franchises prepare orders quicker than others? Use $\alpha = .01$.

## Study Questions

1. A company efficiency expert who wishes to compare the average product assembly times on three different shifts gathers the following data.

Shift 1: 13, 10, 10, 11, 11
Shift 2: 13, 14, 17, 12
Shift 3: 10, 11, 7, 9, 8, 9

where the numbers give observed assembly times (in minutes). Is there sufficient evidence to conclude that the average assembly times on the three shifts are different?

Note that the sample sizes are different, so our procedure must be modified slightly. As before, we calculate the sample means, $\bar{x}_1$, $\bar{x}_2$, and $\bar{x}_3$, and the sample variances, $s_1$, $s_2$, and $s_3$. The overall mean $\mu$ is found by averaging all the sample data together (rather than averaging $\bar{x}_1$, $\bar{x}_2$, and $\bar{x}_3$ as before). To calculate the variance between samples, we use a weighting based on the number of elements in each sample: $n_1$, $n_2$, and $n_3$:

$$\sigma^2_{between} = \frac{n_1(\bar{x}_1 - \mu)^2 + n_2(\bar{x}_2 - \mu)^2 + n_3(\bar{x}_3 - \mu)^2}{k - 1}$$

where $k$, again, is the number of samples. Similarly, the variance within samples must also be based on a weighting to take into consideration the unequal sample sizes:

$$\sigma^2_{within} = \frac{(n_1 - 1)s_1^2 + (n_2 - 1)s_2^2 + (n_3 - 1)s_3^2}{(n_1 + n_2 + n_3) - k}$$

The $F$-ratio is still $\sigma^2_{between} / \sigma^2_{within}$, and the test proceeds as before. Perform this test on assembly line times at the 5% significance level.

2. For Study Question 1, show that if the sample sizes are equal, that is, if $n_1 = n_2 = n_3 = n$, the calculations reduce to those given in the chapter.

# Part 5

# THE POPULATION PROPORTION

---

*Chapters 11–15 dealt with estimates and tests for a population* mean. *Equally important in numerous applications are techniques and procedures involving a population* proportion. *For example, what proportion of registered voters support a proposed bond issue? What proportion of dentists recommend sugarless gum? What proportion of defense contracts are canceled because of cost overruns? What proportion of new houses sell during the first month on the market? Even when more refined measurements are possible, it may be sufficient to simply look at a proportion. For example, it may be necessary to know, not the mean salary, but rather just what proportion of the salaries are above $50,000. It may be necessary to know, not the mean height of basketball players, but rather just the proportion of players above 6 feet 5 inches tall. And it may be necessary to know, not the mean number of new policies sold each week, but rather just the proportion of weeks for which at least ten new policies are sold.*

*Just as with means, it is usually impossible or impracticable to gather complete information about proportions. Thus we must use techniques to make inferences about a population proportion when only a sample proportion is available. As we should now expect, results will involve confidence intervals and degrees of certainty;*

*that is, we cannot make any inference about a population proportion with 100% certainty. However, with procedures very similar to those used to study means, and by making use of techniques developed for working with means, we will be able to draw inferences about proportions. In Chapters 16–18 confidence interval estimates and hypothesis tests for both single-population proportions and differences in two proportions will be examined.*

# 16
# CONFIDENCE INTERVAL OF THE PROPORTION

Whereas the mean is basically a quantitative measurement, the proportion is essentially a qualitative approach. The interest is simply in the presence or absence of some attribute. We count the number of *yes* responses and form a proportion. For example, what proportion of drivers wear seat belts? What proportion of SCUD missiles can be intercepted? What proportion of new stereo sets have a certain defect?

This separation of the population into "haves" and "have-nots" suggests that we can make use of our earlier work on binomial distributions. We also keep in mind that, when *n* (trials, or in this case sample size) is large enough, the binomial can be approximated by the normal.

## DISTRIBUTION OF SAMPLE PROPORTIONS

In this chapter we are interested in estimating a population proportion $\pi$ by considering a single sample proportion $\bar{p}$. This sample proportion is just one of a whole universe of sample proportions, and to judge its significance we must know how sample proportions vary. Consider the set of proportions from all possible samples of a specified size *n*. It seems reasonable that these proportions will cluster around the population proportion, and that the larger the chosen sample size, the tighter will be the clustering.

How do we calculate the mean and variance of the set of population proportions? Suppose the sample size is *n*, and the actual population proportion is $\pi$. From our work on binomial distributions, we remember that the mean and standard deviation for the number of successes in a given sample are $\pi n$ and $\sqrt{\{n\pi (1 - \pi)\}}$ , and for large *n* the complete distribution begins to look "normal."

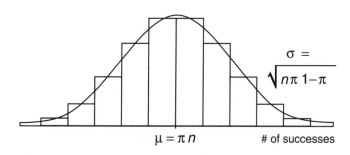

$$\sigma = \sqrt{n\pi \, 1 - \pi}$$

$$\mu = \pi n \qquad \text{\# of successes}$$

Here, however, we are interested in the *proportion*, rather than the *number* of successes. From Chapter 2 we remember that when we multiply or divide every element by a constant, we multiply or divide both the mean and the standard deviation by the same constant. In this case, to change number of successes to proportion of successes, we must divide by *n*:

$$\mu_{\bar{p}} = \frac{\pi n}{n} = \pi \qquad \text{and} \qquad \sigma_{\bar{p}} = \frac{\sqrt{n\pi(1-\pi)}}{n} = \sqrt{\frac{\pi(1-\pi)}{n}}$$

Furthermore, if each element in an approximately normal distribution is divided by the same constant, it is reasonable that the result is still an approximately normal distribution.

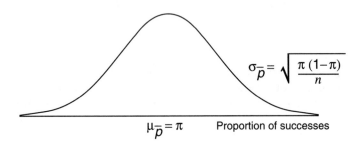

$$\sigma_{\bar{p}} = \sqrt{\frac{\pi(1-\pi)}{n}}$$

$$\mu_{\bar{p}} = \pi \qquad \text{Proportion of successes}$$

Thus the principle forming the basis of much of what we do in the following chapters is as follows:

Start with a population with a given proportion $\pi$. Take all samples of size *n*. Compute the proportion in each of these samples. Then:

1. The set of all sample proportions will be approximately *normally* distributed.
2. The *mean* of the set of sample proportions will equal $\pi$, the population proportion.
3. The *standard deviation* $\sigma_{\bar{p}}$ of the set of sample proportions will be approximately equal to $\sqrt{\dfrac{\pi(1-\pi)}{n}}$.

**Example 16.1**    Suppose that 70% of all dialysis patients will survive for at least 5 years. If 100 new dialysis patients are selected at random, what is the probability that the proportion surviving for at least 5 years will exceed 80%?

*Answer:* The set of sample proportions is approximately normally distributed with mean .70 and standard deviation

$$\sigma_{\bar{p}} = \sqrt{\frac{(.7)(.3)}{100}} = .0458$$

With a $z$-score of $(.80-.70)/.0458 = 2.18$, the probability the sample proportion exceeds 80% is $.5000-.4854 = .0146$.

In finding confidence interval estimates for the population proportion $\pi$, how do we find $\sigma_{\bar{p}} = \sqrt{\dfrac{\pi(1-\pi)}{n}}$ since $\pi$ is unknown? The reasonable procedure is to use the sample proportion $\bar{p}$:

$$\sigma_{\bar{p}} \approx \sqrt{\frac{\bar{p}(1-\bar{p})}{n}}$$

Finally, as with means, if we have a certain confidence that a sample proportion lies within a specified interval around the population proportion, we have the same confidence that the population proportion lies within a specified interval about the sample proportion (the distance from Missoula to Whitefish is the same as the distance from Whitefish to Missoula).

## CONFIDENCE INTERVAL ESTIMATE OF THE POPULATION PROPORTION

**Example 16.2**   If 64% of a sample of 550 people leaving a shopping mall claim to have spent over $25, determine a 99% confidence interval estimate for the proportion of shopping mall customers who spend over $25.

**Answer:** Since $\bar{p} = .64$, the standard deviation of the set of sample proportions is

$$\sigma_{\bar{p}} \approx \sqrt{\frac{(.64)(.36)}{550}} = .0205$$

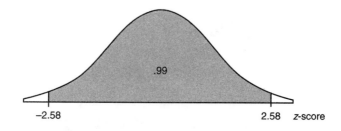

The 99% confidence interval estimate for the population proportion is $.64 \pm 2.58(.0205) = .64 \pm .053$. Thus we are 99% certain that the proportion of shoppers spending over $25 is between .587 and .693.

**Example 16.3** In a random sample of machine parts, 18 of 225 were found to be damaged in shipment. Establish a 95% confidence interval estimate for the proportion of machine parts that are damaged in shipment.

*Answer:* The sample proportion is $\bar{p}$ = 18/225 = .08, and the standard deviation of the set of sample proportions is

$$\sigma_{\bar{p}} \approx \sqrt{\frac{(.08)(.92)}{225}} = .0181$$

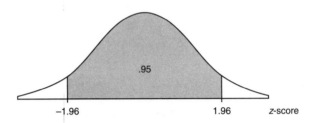

The 95% confidence interval estimate for the population proportion is .08 ± 1.96(.0181) = .08 ± .035. Thus we are 95% certain that the proportion of machine parts damaged in shipment is between .045 and .115.

Suppose there are 50,000 parts in the entire shipment. We can translate from proportions to actual numbers: .045(50,000) = 2250 and .115(50,000) = 5750, so we can be 95% confident that there are between 2250 and 5750 defective parts in the whole shipment.

**Example 16.4** A telephone survey of 1000 adults was taken shortly after the United States began bombing Iraq.
a. If 832 voiced their support for this action, with what confidence can it be asserted that 83.2% ± 3% of the adult U.S. population supported the decision to go to war?

*Answer:*

$$\bar{p} = \frac{832}{1000} = .832, \quad \text{so} \quad \sigma_{\bar{p}} \approx \sqrt{\frac{(.832)(.168)}{1000}} = .0118$$

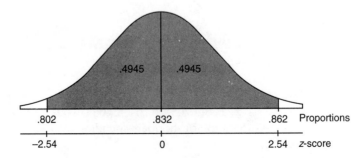

The relevant $z$-scores are $\pm.03/.0118 = \pm2.54$. Table A gives a probability of .4945, so our answer is $2(.4945) = .9890$. In other words, $83.2\% \pm 3\%$ is a 98.90% confidence interval estimate for U.S. adult support of the war decision.

b. If the adult U.S. population is 191,000,000, estimate the actual numerical support.

***Answer:*** Since $.802(191,000,000) \approx 153,000,000$, while $.862(191,000,000) \approx 165,000,000$, we can be 98.90% sure that between 153 and 165 million adults supported the initial bombing decision.

**Example 16.5**  A U.S. Department of Labor survey of 6230 unemployed adults classified people by marital status, sex, and race. The raw numbers are as follows:

|  | white, 16 yrs and over | | | non-white, 16 yrs and over | | |
|---|---|---|---|---|---|---|
|  | **Married** | **Widow/Div** | **Single** | **Married** | **Widow/Div** | **Sing** |
| Men | 1090 | 337 | 1168 | 266 | 135 | 503 |
| Women | 952 | 423 | 632 | 189 | 186 | 349 |

a. Find a 90% confidence interval estimate for the proportion of unemployed men who are married.

***Answer:*** Totaling the first row across, we find that there are 3499 men in the survey; $1090 + 266 = 1356$ of these are married. Therefore

$$\bar{p} = \frac{1356}{3499} = .3875 \quad \text{and}$$

$$\sigma_{\bar{p}} \approx \sqrt{\frac{(.3875)(.6125)}{3499}} = .008236$$

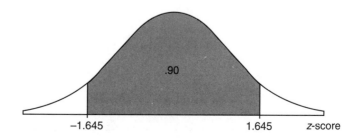

Thus the 90% confidence interval estimate is $.3875 \pm 1.645(.008236) = .3875 \pm .0135$.

b. Find a 98% confidence interval estimate of the proportion of unemployed single persons who are women.

*Answer:* There are 1168 + 632 + 503 + 349 = 2652 singles in the survey. Of these 632 + 349 = 981 are women, so

$$\bar{p} = 981/2652 = .3699 \quad \text{and}$$

$$\sigma_{\bar{p}} \approx \sqrt{\frac{(.3699)(.6301)}{2652}} = .009375$$

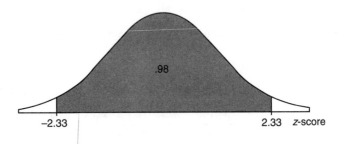

The 98% confidence interval estimate is .3699 ± 2.33(.009375) = .3699 ± .0218.

**Example 16.6**    An assembly-line quality check involves the following procedure: A sample of size 50 is randomly picked, and the machinery is shut down for repairs if the percentage of defective items in the sample is *c* percent or more. Find the value of *c* that results in a 90% chance that the machinery will be stopped if, on the average, the percentage of defective items it is producing is 15%.

*Answer:* Since $\pi = .15$,

$$\sigma_{\bar{p}} \approx \sqrt{\frac{(.15)(.85)}{50}} = .0505$$

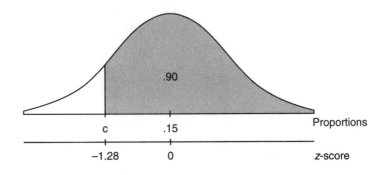

The $z$-score corresponding to a .90 probability is $-1.28$, so $c = .15 - 1.28(.0505) = .085$ or 8.5%.

# SELECTING A SAMPLE SIZE

One important consideration in setting up a survey is the choice of sample size. To obtain a smaller, more precise interval estimate of the population proportion, we must either decrease the degree of confidence or increase the sample size. Similarly, if we want to increase the degree of confidence, we may either accept a wider interval estimate or increase the sample size. Again, while choosing a larger sample size may seem desirable, in the real world this decision involves time and cost considerations.

In setting up a survey to obtain a confidence interval estimate of the population proportion, what should we use for $\sigma_{\hat{p}}$? To answer this question, we first must consider how large $\sqrt{\pi(1 - \pi)}$ can be. Plotting various values of $\pi$:

| $\pi$ | .1 | .2 | .3 | .4 | .5 | .6 | .7 | .8 | .9 |
|---|---|---|---|---|---|---|---|---|---|
| $\sqrt{\pi(1 - \pi)}$ | .3 | .4 | .458 | .490 | .5 | .490 | .458 | .4 | .3 |

gives .5 for the intuitive answer. Thus $\sqrt{\dfrac{\pi(1 - \pi)}{n}}$ is at most $\dfrac{.5}{\sqrt{n}}$ We make use of this fact to determine sample sizes in problems such as Examples 16.7 and 16.8.

**Example 16.7** An EPA investigator wants to know the proportion of fish that are inedible because of chemical pollution downstream of an offending factory. If the answer needs to be within $\pm.03$ at the 96% confidence level, how many fish should be in the sample tested?

*Answer:* We want $2.05\sigma_{\hat{p}} \le .03$. By the above remark, $\sigma_{\hat{p}}$ is at most $.5/\sqrt{n}$, so it is sufficient to consider $2.05(.5/\sqrt{n}) \le .03$. Algebraically, we get $\sqrt{n} \ge 2.05(.5)/.03 = 34.17$, and $n \ge 1167.4$. Therefore, choosing a sample of 1168 fish will give the inedible proportion to within $\pm.03$ at the 96% level.

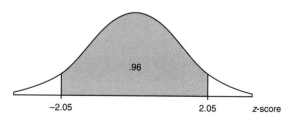

Note that the accuracy of the estimate does *not* depend on what fraction of the whole population we have sampled. What is critical is the *absolute size* of the sample. Is some minimal value of $n$ necessary for the procedures we are using to be meaningful? Since we are using the normal approximation to the binomial, both $np$ and $n(1 - p)$ should be at least 5 (see Chapter 10).

**Example 16.8**    A study is undertaken to determine the proportion of industry executives who believe that workers' pay should be based on individual performance. How many executives should be interviewed if an estimate is desired at the 99% confidence level to within $\pm.06$? To within $\pm.03$? To within $\pm.02$?

**Answer:** Algebraically, $2.58(.5/\sqrt{n}) \leq .06$ gives $\sqrt{n} \geq 2.58(.5)/.06 = 21.5$, so $n \geq 462.25$. Similarly, $2.58(.5/\sqrt{n}) \leq .03$ gives $\sqrt{n} \geq 2.58(.5)/.03 = 43$, so $n \geq 1849$. Finally, $2.58(.5/\sqrt{n}) \leq .02$ gives $\sqrt{n} \geq 2.58(.5)/.02 = 64.5$, so $n \geq 4160.25$. Thus 463, 1849, or 4161 executives should be interviewed, depending on the accuracy desired.

To cut the interval estimate in half (from $\pm.06$ to $\pm.03$), we would have to increase the sample size fourfold (from 462.25 to 1849). To cut the interval estimate to a third (from $\pm.06$ to $\pm.02$), a ninefold increase (from 462.25 to 4160.25) in the sample size would be required.

More generally, to divide the interval estimate by $d$ without affecting the confidence level, we must increase the sample size by a multiple of $d^2$.

# EXERCISES

1. A *USA Today* "*Lifeline*" column reported that in a survey of 500 people 39% said they watch their bread while it's being toasted. Establish a 90% confidence interval estimate for the percentage of people who watch their bread being toasted.

2. In a survey funded by Burroughs-Welcome, 750 of 1000 adult Americans said they didn't believe they could come down with a sexually transmitted disease (STD). Construct a 95% confidence interval estimate for the proportion of adult Americans who don't believe they can catch a STD.

3. During the 1991 recession, a random sample of 130 of a bank's 13,000 home mortgage accounts showed 26 such loans delinquent in payments.
   a. Determine a 99% confidence interval estimate for the proportion of home mortgage accounts that are behind in their payments.
   b. Translate this answer into an estimate of the actual number of delinquent accounts.

4. In a random sampling of 450 hospital deaths, the following numbers were noted with regard to underlying causes.

**Cause of Death**

|  | Heart disease | Cancer | Motor vehicle accidents | other |
|---|---|---|---|---|
| Adult males | 122 | 52 | 18 | 27 |
| Adult females | 95 | 60 | 16 | 15 |
| Children | 8 | 12 | 15 | 10 |

a. Establish a 98% confidence interval estimate for the proportion of all deaths due to heart disease.
b. Establish a 94% confidence interval estimate for the proportion of female deaths due to cancer.
c. Establish an 80% confidence interval estimate for the proportion of motor vehicle deaths which involve children.

5. A 1993 *Los Angeles Times* poll of 1703 adults revealed that only 17% thought the media was doing a "very good" job. With what degree of confidence could the newspaper say that 17% ± 2% of adults believe the media is doing a "very good" job?

6. Suppose that, before marketing a new "improved" laundry detergent, an advertising agency plans to sample 60 consumers and check whether at least $c$% notice an improvement. Find the value of $c$ that results in a 95% chance of acceptance of the detergent if, in reality, 75% of all consumers can detect an improvement.

7. A politician wants to know what percentage of the voters support her position on the issue of forced busing for integration. What size voter sample should be obtained to determine with 90% confidence the support level to within 4%?

8. In a 1994 *New York Times* poll about a candidate's popularity, the newspaper claimed that in 19 of 20 cases its poll results should be no more than 3 percentage points off in either direction.
a. What confidence level are the pollsters working with?
b. What size sample should they have obtained?

## Study Questions
1. Suppose that, of a group of six ($N = 6$) people: $A$, $B$, $C$, $D$, $E$, and $F$, two ($B$ and $E$) own their own houses. Thus the proportion of the group owning their homes is $\pi = 2/6 = 1/3$. Now consider all possible samples of size $n = 4$.
a. List the samples [there are $C(6,4) = 15$], and determine the proportion $\bar{p}$ of home owners for each sample.
b. Calculate the mean and standard deviation for this set of 15 proportions.

Show that the mean is equal to $\pi$, and that the standard deviation is equal to $\sqrt{\dfrac{\pi(1-\pi)}{n}}\sqrt{\dfrac{N-n}{N-1}}$.

c. What happens to $\sqrt{\dfrac{\pi(1-\pi)}{n}}\sqrt{\dfrac{N-n}{N-1}}$ if the population size $N$ is very large?

2. To maximize $\sqrt{\pi(1-\pi)}$, it is sufficient to maximize $\pi(1-\pi) = \pi - \pi^2$. Let $f(\pi) = \pi - \pi^2$, and use elementary calculus to show that $f$ takes its maximum value when $\pi = .5$. Note that the value of $\sqrt{\pi(1-\pi)}$ at $\pi = .5$ is $.5$. Alternatively, graph the parabola $y = \pi - \pi^2$, and note that the vertex is at $\pi = .5$.

3. Pollsters often express results in a form such as 43% plus or minus 5%.
   a. What critical piece of information is missing from such a statement?
   b. How does this omission affect the usefulness, meaning, or strength of such statements?

# 17
# HYPOTHESIS TEST OF THE PROPORTION

Closely related to the problem of estimating a population proportion is the problem of testing a hypothesis about a population proportion. For example, a travel agency might determine an interval estimate for the proportion of sunny days in the Virgin Islands or, alternatively, might test a tourist bureau's claim about the proportion of sunny days. A major stockholder in a construction company might ascertain an interval estimate for the proportion of successful contract bids or, alternatively, might test a company spokesperson's claim about the proportion of successful bids. A social scientist might find an interval estimate for the proportion of homeless children who attend school or, alternatively, might test a school board member's claim about the proportion of such children who are still able to go to classes. In each of these cases, the researcher must decide on whether the interest lies in an interval estimate of a population proportion or in a hypothesis test of a claimed proportion.

The general testing procedure is very similar to that developed for testing a claim about a population mean. Again, there is a specific hypothesis to be tested, called the *null hypothesis*, which is stated in the form of an equality statement about the population proportion (for example, $H_0$: $\pi = .37$). There is the *alternative hypothesis*, stated in the form of an inequality (for example, $H_a$: $\pi < .37$ or $H_a$: $\pi > .37$ or $H_a$: $\pi \neq .37$). The null hypothesis cannot be proved incorrect with absolute certainty; rather, it may be shown to be improbable.

The testing procedure involves picking a sample and comparing the sample proportion $\bar{p}$ to the claimed population proportion $\pi$. A *critical value* (or critical values) $c$ is (are) chosen to gauge the significance of the sample statistic. If the observed $\bar{p}$ is further from the claimed proportion $\pi$ than is the critical value, we say that there is sufficient evidence to reject the null hypothesis. For example, if $H_0$: $\pi < .37$ and if $c = .33$, a sample $\bar{p} = .35$ would not be sufficient evidence to reject $H_0$.

In practice, an acceptable $\alpha$-*risk*, or probability of committing a *Type I error* and mistakenly rejecting a true null hypothesis, is chosen. This $\alpha$-risk, also called the *significance level* of the test, is used to determine the critical value(s). The strength of the sample statistic $\bar{p}$ can further be gauged through its associated *p-value* which gives the smallest value of $\alpha$ for which $H_0$ would be rejected. Finally, the possibility of a mistaken failure to reject a false null hypothesis. This is called a *Type II error* and has associated probability $\beta$. There is a different value of $\beta$ for each possible correct value for the population parameter $\pi$.

# TYPE I ERRORS, CRITICAL VALUES, p-VALUES

**Example 17.1**  A local restaurant owner claims that only 15% of visiting tourists stay for more than 2 days. A chamber of commerce volunteer is sure that the real percentage is higher. He plans to survey 100 tourists and intends to speak up if at least 18 of the tourists stay longer than 2 days. What is the probability of a Type I error?

***Answer:***

$$H_o: \pi = .15$$

$$H_a: \pi > .15$$

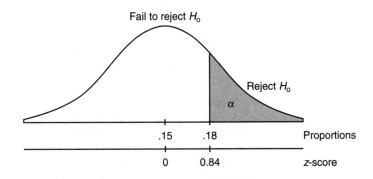

Using the claimed 15%, we calculate the standard deviation of sample proportions to be

$$\sigma_{\bar{p}} = \sqrt{\frac{(.15)(.85)}{100}} = .0357$$

The critical proportion is $c = 18/100 = .18$, so the critical $z$-score is $(.18 - .15)/.0357 = 0.84$. Thus $\alpha = .5000 - .2995 = .2005$. The test, as set up by the volunteer, has a 20.05% chance of mistakenly rejecting a true null hypothesis.

**Example 17.2**  A union spokesperson claims that 75% of union members support a strike if their basic demands are not met. A company negotiator believes the true percentage is lower and runs a hypothesis test at the 10% significance level. What is the conclusion if 87 of 125 union members say they will strike?

**Answer:**

$$H_o: \pi = .75$$
$$H_a: \pi < .75$$

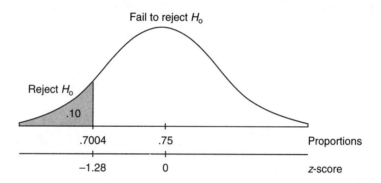

Fail to reject $H_o$

Reject $H_o$

.10

.7004     .75     Proportions

−1.28     0     z-score

We use the claimed proportion to calculate the standard deviation of the sample proportions.

$$\sigma_{\bar{p}} = \sqrt{\frac{(.75)(.25)}{125}} = .03873$$

With $\alpha = .10$ the critical $z$-score is $-1.28$, so the critical proportion is $.75 - 1.28(.03873) = .7004$. The observed sample proportion is $\bar{p} = 87/125 = .696$. Since $.696 < .7004$, there is sufficient evidence to reject $H_o$ at the 10% significance level. The company negotiator should challenge the union claim.

To measure the strength of the disagreement between the sample proportion and the claimed proportion, we calculate the *p-value* (also called the *attained significance level*). The $z$-score for $.696$ is $(.696-.75)/.03873 = -1.39$ with a resulting $p$-value of $.5 - .4177 = .0823$. There is *not* sufficient evidence to reject $H_o$ at the 5%, or even the 8%, significance level.

**Example 17.3**   A cancer research group surveys 500 women over 40 years old to test the hypothesis that 28% of this age group have regularly scheduled mammograms.

a. Should the hypothesis be rejected at the 5% significance level if 151 of the women respond affirmatively?

**Answer:** Since no suspicion is voiced that the 28% claim is low or high, we run a two-sided (also called two-tailed) test.

$$H_o: \pi = .28$$

$$H_a: \pi \neq .28$$

$$\alpha = .05$$

$$\sigma_{\bar{p}} = \sqrt{\frac{(.28)(.72)}{500}} = .0201$$

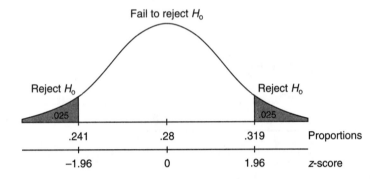

With $\alpha/2 = .025$ in each tail, the critical $z$-scores are $\pm 1.96$, and the critical proportions are $.28 \pm 1.96(.0201) = .241$ and $.319$. The observed $\bar{p} = 151/500 = .302$. Since $.302$ is *not* outside the critical values, there is *not* sufficient evidence to reject $H_o$; that is, the cancer research group should not dispute the 28% claim.

b. What is the *p*-value?

***Answer:*** The $z$-score for $.302$ is $(.302-.28)/.0201 = 1.09$, which corresponds to a probability of $.5000 - .3621 = .1379$. Doubling (because the test is two-sided), we obtain a *p*-value of $2(.1379) = .2758$. Thus, for example, the null hypothesis should not be rejected even if $\alpha$ were a relatively large $.25$.

# TYPE II ERRORS

Why not always choose $\alpha$ to be extremely small so as to eliminate the possibility of mistakenly rejecting a correct null hypothesis? The difficulty is that this choice would simultaneously increase the chance of never rejecting the null hypothesis even if it were far from true. Thus, for a more complete picture, we must also calculate the probability $\beta$ of mistakenly failing to reject a false null hypothesis. As was the case with means, there is a different value $\beta$ for each possible correct value of the population proportion $\pi$. The *operating characteristic curve*, or *OC-curve*, a graphical display of $\beta$ values, is often given in the real-life analysis of a hypothesis test.

**Example 17.4**   A soft-drink manufacturer received a 9% share of the market this past year. The marketing research department plans a telephone survey of 3000 households. If fewer than 8% of respondents indicate they will buy the company's product, the research department will conclude that the company's market share has dropped and will order special new promotions.

a. What is the probability of a Type I error?

***Answer:***

$$H_o: \pi = .09$$

$$H_a: \pi < .09$$

$$\sigma_{\bar{p}} = \sqrt{\frac{(.09)(.91)}{3000}} = .005225$$

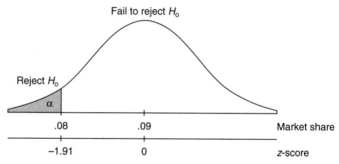

The $z$-score for .08 is $(.08 - .09)/.005225 = -1.91$, so $\alpha = .5 - .4719 = .0281$.

b. What is the probability of a Type II error if the true market share is .085? In other words, if the market share really has dropped to 8.5%, what is the probability that the research department will mistakenly fail to reject the 9% null hypothesis?

***Answer:*** The $z$-score for .08 now is $(.08 - .085)/.005225 = -0.96$. Thus $\beta = .5000 + .3315 = .8315$.

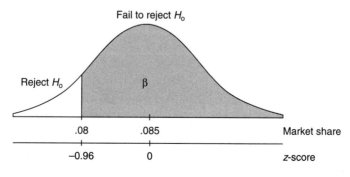

c. What if the true market share is .075?

**Answer:** The $z$-score for .08 now is $(.08 - .075)/.005225 = 0.96$. Thus $\beta = .5000 - .3315 = .1685$.

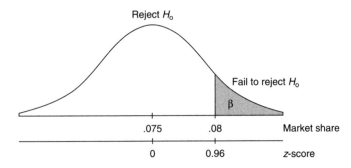

Note that the further the true value is (in the suspected direction) from the claimed null hypothesis, the smaller the probability is of failing to reject the false claim.

**Example 17.5** A building inspector believes that the percentage of new construction with serious code violations may be even greater than the previously claimed 7%. A hypothesis test is planned on 200 new homes at the 10% significance level. What is the value of $\beta$ if the true percentage of new constructions with serious violations is 9%? Is 11%? Is 13%?

**Answer:**

$$H_o: \pi = .07$$

$$H_a: \pi > .07$$

$$\alpha = .10$$

$$\sigma_{\bar{p}} = \sqrt{\frac{(.07)(.93)}{200}} = .018$$

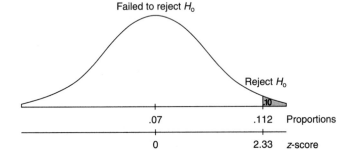

The 10% significance level gives a critical $z$-score of 2.33 and a critical proportion of $.07 + 2.33(.018) = .112$.

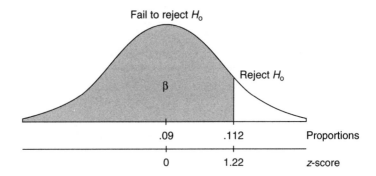

If the true percentage is 9%, the $z$-score for .112 is $(.112 - .09)/.018 = 1.22$. Then $\beta = .5000 + .3888 = .8888$.

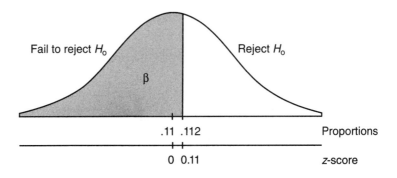

If the true percentage is 11%, the $z$-score for .112 is $(.112 - .11)/.018 = 0.11$. Then $\beta = .5000 + .0438 = .5438$.

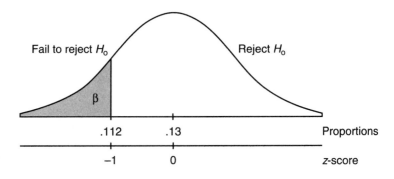

If the true percentage is 13%, the $z$-score for .112 is $(.112 - .13)/.018 = -1$. Then $\beta = .5000 - .3413 = .1587$.

A tabulation of these values (and a few more) for β is as follows:

| True π | .08 | .09 | .10 | .11 | .12 | .13 | .14 | .15 |
|---|---|---|---|---|---|---|---|---|
| β | .9625 | .8888 | .7486 | .5438 | .3300 | .1587 | .0594 | .0174 |

# OPERATING CHARACTERISTIC AND POWER CURVES

The resulting graph is called the *operating characteristic (OC) curve.*

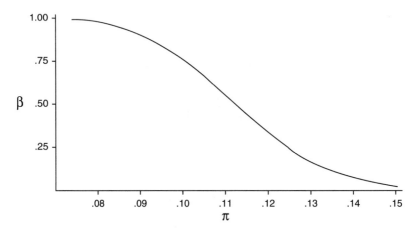

If β is the probability of failing to reject a false null hypothesis, then 1−β is the probability of rejecting the false null hypothesis. The *power* of a hypothesis test is the probability that a Type II error is not committed, and the graph of 1−β, shown below, is called the *power curve.*

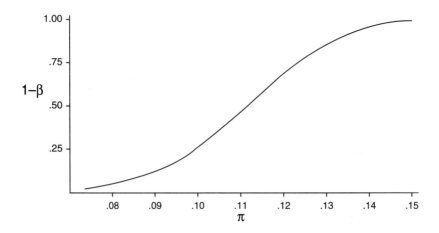

Both the operating characteristic curve and the related power curve help us gauge the effectiveness of a hypothesis test.

# EXERCISES

1. In an effort to curb certain diseases, especially AIDS, San Francisco has a program whereby drug users can exchange used needles for fresh ones. As reported in the *Journal of the American Medical Association* (January 12, 1994), 35% of 5644 intravenous drug users in San Francisco admitted to sharing needles. Is this sufficient evidence at the 1% significance level to say that the rate has dropped from the pre-needle-exchange rate of 66%?

2. In 1993 it was estimated that 1,000,000 of 141,000,000 adults (0.71%) in the United States were HIV positive. A later study, reported in *The New York Times*, found 29 HIV positive adults in a sample of 7782. Is this sufficient evidence to dispute the 0.71% claim at the 5% significance level?

3. The 1948 Kinsey study of human sexuality claimed that 10% of Americans are homosexuals. In a University of Chicago study, reported in *The New York Times*, of 3432 people, approximately 2.1% identified themselves as homosexuals.
   a. What is the *p*-value?
   b. Is this sufficient evidence to reject the Kinsey study at the 1% significance level?

4. According to the *Los Angeles Times*, the president of the National Coffee Association has claimed that "caffeine as normally consumed poses no health risk . . . and cannot be described as a substance of dependence." Researchers from Johns Hopkins University reported that in a sample of 27 coffee drinkers 16 displayed all the signs that define classic dependence.
   a. What is the *p*-value?
   b. Is this sufficient evidence to reject the coffee association president's claim at the 2% significance level?

5. One nationwide report on physical fitness claims that the proportion of middle-level managers who participate in a regular exercise program is .13. A company considering the provision of free health club memberships for its employees plans to survey 200 managers. The new program will be installed immediately if the proportion of exercisers is less than .10.
   a. What is the probability of a Type I error, that is, of rushing ahead when there really is no difference in managers' exercise participation from the nationwide proportion?
   b. What is the probability of a Type II error if the true company proportion is .12? If it is .09?

6. A test is planned at the 10% significance level to determine whether more than 40% of persons eligible for food stamps make full use of the program.

If 500 eligible people are interviewed, determine the probability of a Type II error if the true percentage of users is 42%. What if it is 44%?

7. A business magazine article states that 82% of college graduates majoring in international economics find work in their field. A guidance counselor believes that the true figure is lower and runs a hypothesis test on 230 such graduates at the 5% significance level.
   a. Find the value of $\beta$ for each of the following eight possible true percentages: 83%, 84%, . . . , 90%.
   b. Sketch the operating characteristic curve.

## Study Questions

1. Suppose that your rival in the marketing office has what you consider to be a dumb idea. Your boss, who knows nothing about statistics, is relatively conservative but will follow up on this dumb idea if there is good evidence that more than 25% of the company's customers like it. A survey of 40 customers is taken, and 12 of them favor the idea. Your rival points out that $12/40 = 30\%$. You know a little about statistics and perform the usual hypothesis test with $H_0$: $\pi = .25$ and $H_1$: $\pi > .25$. Write a paragraph explaining to your boss why these data are *not* strong evidence that more than 25% of the customers like your rival's idea.

2. (*Type II error in a two-tailed test*) It is estimated that 60% of all grocery store customers clip, save, and use advertisers' money-off coupons. A store manager surveys 300 customers and runs a test at the 5% significance level. What is the value of $\beta$ if the true percentage of customers using coupons is 53%? 58%? 63%? 68%?

# 18
# DIFFERENCES IN POPULATION PROPORTIONS

Numerous important and interesting applications of statistics involve the comparison of two *population proportions*. For example, is the proportion of satisfied purchasers of American automobiles greater than that for buyers of Japanese cars? How does the percentage of surgeons recommending a new cancer treatment compare with the corresponding percentage of oncologists? What can be said about the difference between the proportion of single parents who are on welfare and the proportion of two-parent families on welfare?

Our procedure involves comparing two sample proportions. When is a difference between two such sample proportions significant? We note that we are dealing with one difference from the set of all possible differences obtained by subtracting sample proportions of one population from sample proportions of a second population. To judge the significance of one particular difference, we must first determine how the differences vary among themselves. As in the similar discussion regarding the comparison of two means, a necessary tool from Chapter 2 is that the variance of a set of differences is equal to the sum of the variances of the individual sets; that is,

$$\sigma_d^2 = \sigma_1^2 + \sigma_2^2$$

Now if $\sigma_1 = \sqrt{\dfrac{\pi_1(1 - \pi_1)}{n_1}}$ and $\sigma_2 = \sqrt{\dfrac{\pi_2(1 - \pi_2)}{n_2}}$ then

$$\sigma_d^2 = \frac{\pi_1(1 - \pi_1)}{n_1} + \frac{\pi_2(1 - \pi_2)}{n_2} \quad \text{and}$$

$$\sigma_d = \sqrt{\frac{\pi_1(1 - \pi_1)}{n_1} + \frac{\pi_2(1 - \pi_2)}{n_2}}$$

As in the preceding chapters on proportions, we are using the normal to estimate the binomial, so we will assume that

$$n_1\pi_1, \ n_1(1 - \pi_1), \ n_2\pi_2, \quad and \ n_2(1 - \pi_2)$$

are all at least 5.

# CONFIDENCE INTERVAL ESTIMATES FOR DIFFERENCES

Often it is clear that some difference exists between two population proportions, and we would like to find a numerical measure of the difference. Using samples, we cannot determine this difference *exactly*, however, we can say, with a specified degree of confidence, that the difference lies in a certain interval. We follow the same procedure as set forth in Chapter 16, this time using

$$\pi_1 - \pi_2, \quad \sqrt{\frac{\pi_1(1 - \pi_1)}{n_1} + \frac{\pi_2(1 - \pi_2)}{\pi_2}}, \quad and \quad \bar{p}_1 - \bar{p}_2$$

$$in\ place\ in\ \pi, \quad \sqrt{\frac{\pi(1 - \pi)}{n}}, \quad and \quad \bar{p}$$

**Example 18.1**   Suppose that 84% of a sample of 125 nurses working 7 A.M. to 3 P.M. shifts in city hospitals express positive job satisfaction, while only 72% of a sample of 150 nurses on 11 P.M. to 7 A.M. shifts express similar fulfillment. Establish a 90% confidence interval estimate for the difference.

***Answer:***

$$n_1 = 125,\ n_2 = 150$$

$$\bar{p}_1 = .84,\ \bar{p}_2 = .72$$

$$\sigma_d \approx \sqrt{\frac{(.84)(.16)}{125} + \frac{(.72)(.28)}{150}} = .0492$$

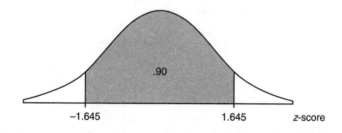

The observed difference is $.84 - .72 = .12$, and the critical $z$-scores are $\pm 1.645$. The confidence interval estimate is $.12 \pm 1.645(.0492) = .12 \pm .081$. We can be 90% certain that the proportion of satisfied nurses on 7 to 3 shifts is between .039 and .201 higher than for the proportion for nurses on 11 to 7 shifts.

**Example 18.2**  A grocery store manager notes that, in a sample of 85 people going through the "under seven items" checkout line, only 10 paid with checks; whereas, in a sample of 92 customers passing through the regular line, 37 paid with checks. Find a 95% confidence interval estimate for the difference between the proportion of customers going through the two different lines who use checks.

***Answer:***

$$n_1 = 85, \qquad n_2 = 92$$

$$\bar{p}_1 = \frac{10}{85} = .118$$

$$\bar{p}_2 = \frac{37}{92} = .402$$

$$\sigma_d \approx \sqrt{\frac{(.118)(.882)}{85} + \frac{(.402)(.598)}{92}} = .0619$$

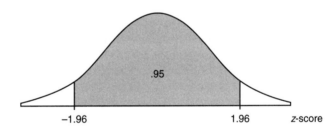

The observed difference is $.118 - .402 = -.284$, and the critical $z$-scores are $\pm 1.96$. Thus, the confidence interval estimate is $-.284 \pm 1.96(.0619) = -.284 \pm .121$. The manager can be 90% sure that the proportion of customers passing through the "under seven items" line who use checks is between .163 and .405 lower than the proportion going through the regular line who use checks.

# HYPOTHESIS TESTS

For many problems the null hypothesis states that the population proportions are equal, or, equivalently, that their difference is 0:

$$H_0: \pi_1 - \pi_2 = 0$$

The alternative hypothesis is then

$$H_a: \pi_1 - \pi_2 < 0, \ H_a: \pi_1 - \pi_2 > 0, \ \text{or} \ H_a: \pi_1 - \pi_2 \neq 0,$$

where the first two possibilities lead to one-sided (one-tailed) tests, while the third possibility leads to two-sided (two-tailed) tests.

Since the null hypothesis is that $\pi_1 = \pi_2$, we call this common value $\pi$, and use it in calculating $\sigma_d$:

$$\sigma_d = \sqrt{\frac{\pi(1-\pi)}{n_1} + \frac{\pi(1-\pi)}{n_2}} = \sqrt{\pi(1-\pi)\left(\frac{1}{n_1} + \frac{1}{n_2}\right)}$$

In practice, if

$$\bar{p}_1 = \frac{x_1}{n_1} \quad \text{and} \quad \bar{p}_2 = \frac{x_2}{n_2}$$

we use

$$\bar{p} = \frac{x_1 + x_2}{n_1 + n_2}$$

as an estimate for $\pi$ in calculating $\sigma_d$.

**Example 18.3**   Suppose that, early in an election campaign, a telephone poll of 800 registered voters shows 460 in favor of a particular candidate. Just before election day, a second poll shows only 520 of 1000 registered voters expressing the same preference. At the 10% significance level is there sufficient evidence that the candidate's popularity has decreased?

***Answer:***

$$H_o: \pi_1 - \pi_2 = 0 \qquad \bar{p}_1 = \frac{460}{800} = .575$$

$$H_a: \pi_1 - \pi_2 > 0$$

$$\alpha = .10 \qquad \bar{p}_2 = \frac{520}{1000} = .520$$

$$\bar{p} = \frac{460 + 520}{800 + 1000} = .544$$

$$\sigma_d \approx \sqrt{(.544)(.456)\left(\frac{1}{800} + \frac{1}{1000}\right)} = .0236$$

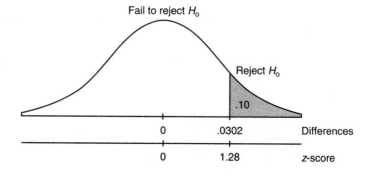

With $\alpha = .10$, the critical $z$-score is 1.28, so the critical difference is $0 + 1.28(.0236) = .0302$. The observed difference is $.575 - .520 = .055$. Since $.055 > .0302$, we conclude that at the 10% significance level the candidate's popularity *has* dropped.

The strength of this drop can be further measured by calculating the $p$-value. The $z$-score for $.055$ is $(.055-0)/.0236 = 2.33$, so the $p$-value is $.5000 - .4901 = .0099$. Thus there is sufficient evidence of a drop in popularity even at the 1% significance level.

**Example 18.4** An automobile manufacturer tries two distinct assembly procedures. In a sample of 350 cars coming off the line using the first procedure, there are 28 with major defects, while a sample of 500 autos from the second line shows 32 with defects.

a. Is the difference significant at the 6% significance level?

***Answer:*** Since there is no mention that one procedure is believed to be better or worse than the other, this is a two-sided test.

$$H_0: \pi_1 - \pi_2 = 0 \qquad \bar{p}_1 = \frac{28}{350} = .080$$

$$H_a: \pi_1 - \pi_2 \neq 0$$

$$\alpha = .06 \qquad \bar{p}_2 = \frac{32}{500} = .064$$

$$\bar{p} = \frac{28 + 32}{350 + 500} = .0706$$

$$\sigma_d \approx \sqrt{(.0706)(.9294)\left(\frac{1}{350} + \frac{1}{500}\right)} = .0179$$

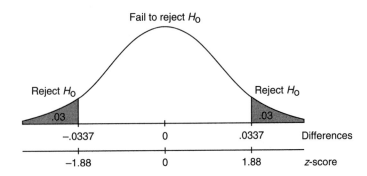

With $\alpha = .06$, we put $.03$ in each tail; the critical $z$-scores are $\pm 1.88$. The critical differences are $0 \pm 1.88(.0179) = \pm.0337$. The observed difference is $.080 - .064 = .016$. Since $.016$ is between the two critical values, we conclude that the observed difference is *not* significant at the 6% level.

b. What is the *p*-value?

*Answer:* The *z*-score for .016 is $(0 - .016)/.0179 = 0.89$, which corresponds to a tail probability of $.5000 - .3133 = .1867$. Doubling because the test is two-sided results in a *p*-value of $2(.1867) = .3734$. Thus the smallest significance level for which the observed difference is significant is a very large 37.34%.

**Example 18.5**   A survey of 5000 medical students compared the career goals of men and women.

| | **Career Goal** | | | | |
|---|---|---|---|---|---|
| | **Surgery** | **Gynecology** | **Pediatrics** | **Psychiatry** | **Other** |
| Men | 312 | 520 | 472 | 610 | 1026 |
| Women | 128 | 350 | 391 | 400 | 791 |

a. Establish a 95% confidence interval estimate for the difference between the proportions of men and women intending to become surgeons.

*Answer:*

$$\bar{p}_1 = \frac{312}{312 + 520 + 472 + 610 + 1026} = \frac{312}{2940} = .106$$

$$\bar{p}_2 = \frac{128}{128 + 350 + 391 + 400 + 791} = \frac{128}{2060} = .062$$

$$\sigma_d \approx \sqrt{\frac{(.106)(.894)}{2940} + \frac{(.062)(.938)}{2060}} = .00778$$

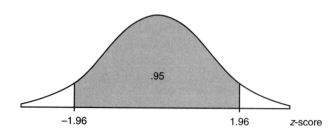

The observed difference is $.106 - .062 = .044$, so the confidence interval estimate is $.044 \pm 1.96(.00778) = .044 \pm .015$. Therefore we can be 95% certain that the proportion of men heading toward a career in surgery is between .029 and .059 higher than the proportion of women.

b. Is the observed difference between the proportions of men and women hoping to become gynecologists significant at the 2% significance level?

***Answer:***

$$H_o: \pi_1 - \pi_2 = 0 \qquad \bar{p}_1 = \frac{520}{2940} = .177$$

$$H_a: \pi_1 - \pi_2 \neq 0$$

$$\alpha = .02 \qquad \bar{p}_2 = \frac{350}{2060} = .170$$

$$\bar{p} = \frac{(520 + 350)}{(5000)} = .174$$

$$\sigma_d \approx \sqrt{(.174)(.826)\left(\frac{1}{2940} + \frac{1}{2060}\right)} = .0109$$

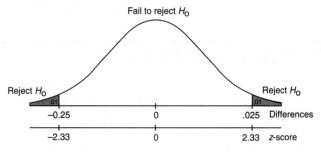

The critical $z$-scores are $\pm 2.33$, so the critical differences are $0 \pm 2.33(.0109) = \pm.025$. The observed difference is $.177 - .170 = .007$. Since $.007$ is between $-.025$ and $.025$, there is *not* sufficient evidence to reject $H_o$. The observed difference between the proportions of men and women with gynecology as a career goal is not significant at the 2% level.

c. Is there sufficient evidence at the 0.5% significance level that a higher proportion of women want to become pediatricians than do men?

***Answer:***

$$H_o: \pi_1 - \pi_2 = 0 \qquad \bar{p}_1 = \frac{391}{2060} = .190$$

$$H_a: \pi_1 - \pi_2 > 0$$

$$\alpha = .005 \qquad \bar{p}_2 = \frac{472}{2940} = .161$$

$$\bar{p} = \frac{391 + 472}{(5000)} = .173$$

$$\sigma_d \approx \sqrt{(.173)(.827)\left(\frac{1}{2060} + \frac{1}{2940}\right)} = .0109$$

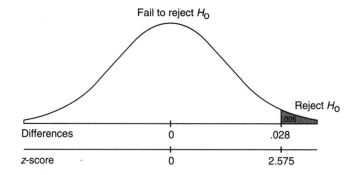

Fail to reject $H_0$

Reject $H_0$

.005

| Differences | 0 | .028 |

| z-score | 0 | 2.575 |

The critical $z$-score is 2.575, so the critical difference is $0 + 2.575(.0109) = .028$. The observed difference is $.190 - .161 = .029$. Since $.029 > .028$, there is sufficient evidence to reject $H_0$. At the 0.5% significance level there is evidence that a greater proportion of women than men plan to become pediatricians.

## SELECTING A SAMPLE SIZE

Setting up experiments or surveys involves many considerations, one of which is *sample size*. For example, one topic in this chapter involves making interval estimates of differences in population proportions. Generally, if we want smaller, more precise interval estimates, we either decrease the degree of confidence *or* increase the sample size. Similarly, if we want to increase the degree of confidence, we may either accept a wider interval *or* again increase the sample size.

In Chapter 16 we noted that $\sqrt{\pi(1 - \pi)}$ is at most .5. Thus

$$\sqrt{\pi(1 - \pi)\left(\frac{1}{n_1} + \frac{1}{n_2}\right)} \leq (.5)\sqrt{\frac{1}{n_1} + \frac{1}{n_2}}$$

Now, if we simplify by insisting that $n_1 = n_2 = n$, the above statement reduces as follows:

$$(.5)\sqrt{\frac{1}{n} + \frac{1}{n}} = (.5)\sqrt{\frac{2}{n}} = \frac{.5\sqrt{2}}{\sqrt{n}}$$

**Example 18.6**   A pollster wants to determine the difference between the proportions of high-income voters and low-income voters who support a decrease in capital gains taxes. If the answer needs to be known to within $\pm.02$ at the 95% confidence level, what size samples should be taken?

***Answer:*** Assuming we will pick the same size samples for the two sample proportions, we have $\sigma_d \leq .5\sqrt{2}/\sqrt{n}$ and $1.96\sigma_d \leq .02$.

Thus, $1.96(.5)\sqrt{2}/\sqrt{n} \leq .02$. Algebraically we find that $\sqrt{n} \geq 1.96(.5)\sqrt{2}/.02 = 69.3$. Therefore, $n \geq 69.3^2 = 4802.5$, and the pollster should use 4803 people for each sample.

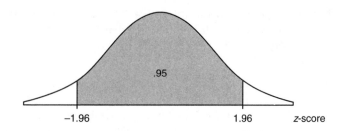

# EXERCISES

1. The National Research Council of the Philippines reported (*Science*, March 11, 1994, page 1491) that 210 of 361 members in biology are women, but only 34 of 86 members in mathematics are women. Establish a 95% confidence interval estimate for the difference in proportions between women in biology and women in mathematics in the Philippines.

2. In a random sample of 300 elderly men, 65% were married, while in a similar sample of 400 elderly women, 48% were married. Determine a 99% confidence interval estimate for the difference between the percentages of elderly men and women who are married.

3. Suppose that 23 of 94 oil company executives recommend selling short on major oil stocks, while 34 of 80 stockbrokers recommend such sales. Construct a 90% confidence interval estimate for the difference between the percentages of oil executives and stockbrokers who recommend selling oil stocks short.

4. The Baldus-Woodworth 5 year study of Georgia's death penalty system noted that 50 of 228 black defendants received the death penalty after conviction for murdering a white person, while 2 of 64 white defendants received the death penalty for murdering a black person. Is there sufficient evidence at the 5% significance level to say that a higher proportion of blacks than whites receive the death penalty in Georgia when the victim is of the other race?

5. Tversky and Gilovich, in their study on the "Hot Hand" in basketball (*Chance*, Winter 1989, page 20), found that Larry Bird hit a second free throw 48 of 53 attempts after the first free throw was missed, and hit a second free throw 251 of 285 attempts after the first free throw was made. Is there sufficient evidence at the 5% significance level to say that the probability Bird makes a second free throw is different depending on whether or not he made the first free throw?

6. The results of a double-blind study on whether AZT taken by pregnant HIV-positive women helps prevent the virus from being passed to the fetus gave seemingly strong results (*New England Journal of Medicine*, November 1994). Of 240 women who took the drug, 8.3% of the children were infected with HIV, while of 240 women given a placebo, 25.5% of the children were infected. Is there sufficient evidence at the 1% significance level to say that AZT helps prevent the HIV virus from being passed from mother to child?

7. *The New York Times* (April 2, 1993, page A12) reported that in a double-blind study of newly diagnosed HIV patients, 92% of 875 patients treated with AZT survived the initial 3 years, while 93% of 875 patients given a placebo were alive 3 years after the diagnosis. Is the difference in survival rates statistically significant?

8. A government survey of income levels of men and women yielded the following numbers for the 7689 person random sampling.

|  | **Earning Group** | | | |
| --- | --- | --- | --- | --- |
|  | **Under $10,000** | **$10-20,000** | **$20-50,000** | **Over $50,000** |
| Men | 383 | 1177 | 2599 | 551 |
| Women | 473 | 1371 | 1084 | 51 |

a. Establish a 96% confidence interval estimate for the difference between the proportions of men and women who earn between $20,000 and $50,000.

b. Construct a 98% confidence interval estimate for the difference between the proportions of persons earning over $50,000 and those earning under $10,000 who are men.

c. Test the claim that a greater proportion of women than men earn under $10,000 at the .5% significance level.

d. Is there a significant difference between the proportions of men and women earning between $10,000 and $50,000 at the 2.5% significance level?

9. A researcher plans to investigate the difference between the proportion of psychiatrists and the proportion of psychologists who believe that most emotional problems have their root causes in patients' childhoods. How large a sample should be taken (same number for each group) so as to be 90% certain of knowing the difference to within $\pm.03$?

## Study Questions

1. The hypothesis tests in this chapter all began with the null hypothesis $H_o$: $\pi_1 - \pi_2 = 0$. However, we can easily modify our procedure to handle $H_o$: $\pi_1 - \pi_2 = k$ for some nonzero constant $k$. In this case we simply use

$$\sigma_d = \sqrt{\frac{\bar{p}_1(1-\bar{p}_1)}{n_1} + \frac{\bar{p}_2(1-\bar{p}_2)}{n_2}}$$ as we did for confidence intervals. For example, suppose that 210 of 3600 alumni responded to a mail solicitation, while 476 of 2700 responded to a telephone appeal for funds. Does this support the claim that the proportion of successful telephone appeals exceeds the proportion of successful mail solicitations by more than .10? Test at the 2.5% significance level.

2. Three in-store surveys are taken: one before, one during, and one after a special promotion for tropical fruits. The proportions of customers expressing interest in these products are 29 of 157, 37 of 165, and 40 of 152 for the surveys before, during, and after the promotion, respectively.
   a. At the 10% significance level, is the interest increase from before to during the promotion significant? From during to after the promotion? From before to after the promotion?
   b. What does the result obtained in part *a* say about the possibility that $\bar{p}_1$ and $\bar{p}_3$ are significantly different when $\bar{p}_1$ and $\bar{p}_2$ are not, and $\bar{p}_2$ and $\bar{p}_3$ are not?

3. If

$$\bar{p}_1 = \frac{x_1}{n_1}, \qquad \bar{p}_2 = \frac{x_2}{n_2}, \qquad \text{and} \quad \bar{p} = \frac{x_1 + x_2}{n_1 + n_2},$$

is it ever true that

$$\bar{p} = \frac{\bar{p}_1 + \bar{p}_2}{2}?$$

[*Hint*: Make a reasonable guess about a simple relationship between $n_1$ and $n_2$, see whether $\bar{p} = \frac{1}{2}(\bar{p}_1 + \bar{p}_2)$ holds true in this case when you plug in some arbitrary numerical values, and then try to show why this equation is true in general.]

# *Part* **6**

# CHI-SQUARE ANALYSIS

*In Chapters 19 and 20 we consider two types of problems involving similar analyses. The first concerns whether or not some observed distributional outcome fits some previously specified pattern. Perfect fits very rarely exist, and we must develop a procedure for measuring the significance of a loose fit. The second type of problem concerns whether two variables are independent or have some relationship. Here we use the same kind of measurement as for the first type to judge the significance of a loose fit with the theoretical pattern based on independence. The kinds and the variety of problems we will be considering are indicated by the following examples.*

*A geneticist might test whether inherited traits can be explained by a* binomial distribution. *A physicist might look to a* Poisson distribution *to further understand alpha particle emissions. A psychologist might study human intelligence patterns in terms of a* normal distribution. *In these and many other applications, the question arises as to how well observed data fit the pattern expected from some specified distribution.*

*A pollster might look at whether or not voter support for a candidate is* independent *of the voters' ethnic backgrounds. An efficiency expert might examine whether or not the likelihood of an accident is* independent *of the shift on which an employee works. A psychologist might test whether or not certain mental abnormalities are* independent *of socioeconomic background.*

*As can be seen, the procedure described in the Chapters 19 and 20 has a wide range of applications.* Chi-square analysis *is one of the most useful statistical techniques, both in cases where other tests are not applicable, and in cases where other tests may be unnecessarily complicated.*

# 19
# CHI-SQUARE TEST FOR GOODNESS-OF-FIT

Among the many theoretical distributions analyzed by statisticians are the binomial, the Poisson, the normal, and the Student-$t$, all of which we have made use of. In addition, numerous other distributions are expected or looked for in specific real-world situations. The critical question is whether or not an observed pattern of data fits some given distribution. A perfect fit cannot be expected, so we must be able to look at discrepancies and make judgments as to the *goodness-of-fit*.

Our approach is similar to that developed earlier. There is the null hypothesis of a good fit, that is, the hypothesis that a given theoretical distribution correctly describes the situation, problem, or activity under consideration. Our observed data consist of one possible sample from a whole universe of possible samples. We ask about the chance of obtaining a sample with the observed discrepancies if the null hypothesis is really true. Finally, if the chance is too small, we reject the null hypothesis and say that the fit is not a good one.

How do we decide about the significance of observed discrepancies? It should come as no surprise that the best information is obtained from squaring the discrepancy values, as this has been our technique for studying variances from the beginning. Furthermore, since, for example, an observed difference of 23 is more significant if the original values are 105 and 128 than if they are 10,602 and 10,625, we must appropriately *weight* each difference. Such weighting is accomplished by dividing each difference by the expected value. The sum of these weighted differences or discrepancies is called *chi-square* and is denoted as $\chi^2$ ($\chi$ is the lower-case Greek letter chi):

$$\chi^2 = \Sigma \frac{(\text{obs} - \text{exp})^2}{\text{exp}}$$

The smaller the resulting $\chi^2$-value, the better is the fit. If the $\chi^2$-value exceeds some critical value, we say there is sufficient evidence to reject the null hypothesis and to claim that the fit is poor.

To decide how large a calculated $\chi^2$-value must be to be significant, that is, to choose a critical value, we must understand how $\chi^2$-values are distributed. A $\chi^2$-distribution is not symmetrical and is always skewed to the right. Just as with the $t$-distribution, there are distinct $\chi^2$-distributions each with an associated df value (number of degrees of freedom). The larger the df value, the closer the $\chi^2$-distribution is to a normal distribution. Note, for example, that squaring the often used $z$-scores 1.645, 1.96, and 2.576 results in 2.706, 3.841, and 6.635, which are entries found in the first row of Table D, the $\chi^2$-distribution table, in the Appendix.

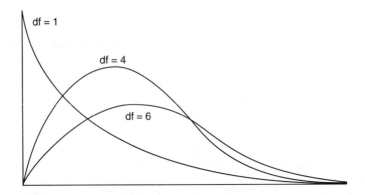

df = 1

df = 4

df = 6

A large value of $\chi^2$ may or may not be significant—the answer depends on which $\chi^2$-distribution we are using. As with the t-distribution, in Table D in the Appendix we give critical $\chi^2$-values for only the more commonly used percentages or probabilities. To use the $\chi^2$-distribution for approximations in goodness-of-fit problems, the individual expected values cannot be too small. An often-used rule of thumb is that no expected value should be less than 5.

# GOODNESS-OF-FIT TEST FOR PRIOR DISTRIBUTION

**Example 19.1** In a recent year, at the 6 P.M. time slot, television channels 2, 3, 4, and 5 captured the entire audience with 30%, 25%, 20%, and 25%, respectively. During the first week of the next season, 500 viewers are interviewed.

a. If viewer preferences have not changed, what number of persons is expected to watch each channel?

***Answer:*** .30(500) = 150, .25(500) = 125, .20(500) = 100, and .25(500) = 125, so we have these results:

Channel

| | 2 | 3 | 4 | 5 |
|---|---|---|---|---|
| Expected number | 150 | 125 | 100 | 125 |

b. Suppose that the actual observed numbers are as follows:

| | Channel | | | |
|---|---|---|---|---|
| | 2 | 3 | 4 | 5 |
| Observed number | 139 | 138 | 112 | 111 |

Do these numbers indicate a change? Are the differences signifi-cant?

***Answer:*** We calculate:

$$\chi^2 = \Sigma\frac{(\text{obs} - \text{exp})^2}{\text{exp}}$$
$$= \frac{(139 - 150)^2}{150} + \frac{(138 - 125)^2}{125} + \frac{(112 - 100)^2}{100}$$
$$+ \frac{(111 - 125)^2}{125} = 5.167$$

Is 5.167 large enough for us to reject the null hypothesis of a good fit between the observed and the expected numbers? To use Table D in the Appendix, we must decide upon an α-risk and then calculate the *df* value. The number of degrees of freedom for this problem is 4 − 1 = 3. (Note that, while the observed values of 139, 138, and 112 can be freely chosen, the fourth value, 111, is absolutely determined because the total must be 500; thus *df* = 3.)

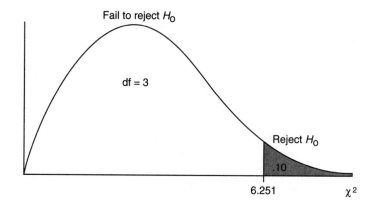

Thus, for example, at a 10% significance level, with *df* = 3, the critical $\chi^2$ is 6.251. Since 5.167 < 6.251, there is *not* sufficient evidence to reject $H_o$.

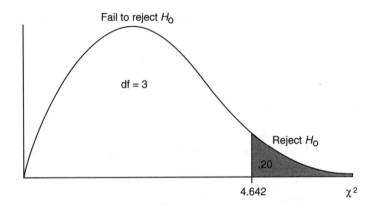

However, with $df = 3$ and a 20% significance level, the critical $\chi^2$ is 4.642. Since $5.167 > 4.642$, there *is* sufficient evidence to reject $H_o$. At the 20% level we would conclude that the fit is not good, and that viewer preferences have changed.

**Example 19.2**  A geneticist claims that four species of fruit flies should appear in the ratio 1:3:3:9. Suppose that a sample of 4000 flies contained 226, 764, 733, and 2277 flies of each species, respectively. At the 10% significance level, is there sufficient evidence to reject the geneticist's claim?

***Answer:*** Since $1 + 3 + 3 + 9 = 16$, according to the geneticists the expected number of fruit flies of each species is as follows:

$$\frac{1}{16}(4000) = 250, \qquad \frac{3}{16}(4000) = 750,$$

$$\frac{3}{16}(4000) = 750, \qquad \frac{9}{16}(4000) = 2250$$

We calculate chi-square:

$$\chi^2 = \frac{(226 - 250)^2}{250} + \frac{(764 - 750)^2}{750} + \frac{(733 - 750)^2}{750}$$
$$+ \frac{(2277 - 2250)^2}{2250} = 3.27$$

With $4 - 1 = 3$ degrees of freedom and $\alpha = .10$, the critical $\chi^2$ value is 6.251. Since $3.27 < 6.251$, there is *not* sufficient evidence to reject $H_o$. At the 10% significance level, therefore, the geneticist's claim should not be rejected.

# GOODNESS-OF-FIT TEST FOR UNIFORM DISTRIBUTION

**Example 19.3**  A grocery store manager wishes to determine whether a certain product will sell equally well in any of five locations in the store. Five displays are set up, one in each location, and the resulting numbers of the product sold are noted.

|  | Location |  |  |  |  |
| --- | --- | --- | --- | --- | --- |
|  | 1 | 2 | 3 | 4 | 5 |
| Actual number sold | 43 | 29 | 52 | 34 | 48 |

Is there enough evidence that location makes a difference? Test at both the 5% and 10% significance levels.

***Answer:*** A total of $43 + 29 + 52 + 34 + 48 = 206$ units were sold. If location doesn't matter, we would expect $206/5 = 41.2$ units sold per location (uniform distribution).

|  | Location |  |  |  |  |
| --- | --- | --- | --- | --- | --- |
|  | 1 | 2 | 3 | 4 | 5 |
| Expected number sold | 41.2 | 41.2 | 41.2 | 41.2 | 41.2 |

$$\text{Thus } \chi^2 = \frac{(43 - 41.2)^2}{41.2} + \frac{(29 - 41.2)^2}{41.2} + \frac{(52 - 41.2)^2}{41.2}$$
$$+ \frac{(34 - 41.2)^2}{41.2} + \frac{(48 - 41.2)^2}{41.2} = 8.903$$

As in the previous examples, the number of degrees of freedom is the number of classes minus 1; that is, df $= 5 - 1 = 4$.

$H_0$: good fit to uniform distribution

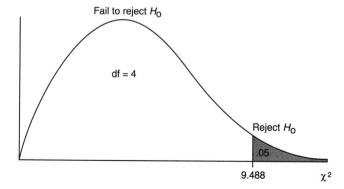

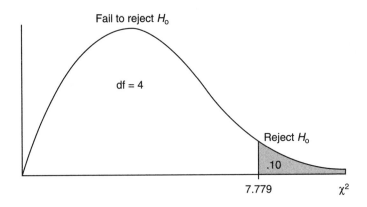

At the 5% level, with df $= 4$, the critical $\chi^2$-value is 9.488, while at the 10% level the critical value is 7.779. Since $8.903 < 9.488$ but $8.903 > 7.779$, there is sufficient evidence to reject $H_o$ at the 10% level, but not at the 5% level. If the grocery store manager is willing to accept a 10% chance of committing a Type I error, then there is enough evidence to claim that location makes a difference.

**Example 19.4**  A pet-food manufacturer runs an experiment to determine whether three brands of dog food are equally preferred by dogs. In the experiment, 150 dogs are individually presented with three dishes of food, each containing a different brand, and their choices are noted. Tabulations show that 62 dogs go to brand $A$, 43 to brand $B$, and 45 to brand $C$. Is there sufficient evidence to say that dogs have preferences among the brands? Test at the 2.5% significance level.

***Answer:*** If dogs have no preferences among the three brands, we would expect $150/3 = 50$ dogs to go to each dish. Thus:

$$\chi^2 = \frac{(62-50)^2}{50} + \frac{(43-50)^2}{50} + \frac{(45-50)^2}{50} = 4.36$$

With $3 - 1 = 2$ degrees of freedom and $\alpha = .025$, the critical $\chi^2$-value is 7.378. Since $4.36 < 7.378$, there is *not* sufficient evidence to say that dogs have preferences among the three given brands. (Of course the company that manufactures brand $A$ will continue to claim that more dogs prefer its food!)

# GOODNESS-OF-FIT TEST FOR BINOMIAL DISTRIBUTION

**Example 19.5**  Suppose that a commercial is run once on television, once on the radio, and once in a newspaper. The advertising agency believes that any potential consumer has a 20% chance of seeing the ad on TV, a 20% chance of hearing it on the radio, and a 20% chance of reading it in the paper. In a telephone survey of 800 consumers, the numbers claiming to have been exposed to the ad 0, 1, 2, or 3 times are as follows.

Exposures to Ad

|  | 0 | 1 | 2 | 3 |
|---|---|---|---|---|
| Observed number of people | 434 | 329 | 35 | 2 |

At the 1% significance level, test the null hypothesis that the number of times any consumer saw the ad follows a binomial distribution with $\pi = .2$.

***Answer:*** The complete binomial distribution with $\pi = .2$ and $n = 3$ is as follows:

$$P(0) = (.8)^3 \qquad\quad = .512$$

$$P(1) = 3(.2)(.8)^2 = .384$$

$$P(2) = 3(.2)^2(.8) = .096$$

$$P(3) = (.2)^3 \qquad\quad = .008$$

Multiplying each of these probabilities by 800 gives the expected number of occurrences: $.512(800) = 409.6$, $.384(800) = 307.2$, $.096(800) = 76.8$, $.008(800) = 6.4$

Exposures to Ad

|  | 0 | 1 | 2 | 3 |
|---|---|---|---|---|
| Expected number of people | 409.6 | 307.2 | 76.8 | 6.4 |

Thus

$$\chi^2 = \frac{(434 - 409.6)^2}{409.6} + \frac{(329 - 307.2)^2}{307.2} + \frac{(35 - 76.8)^2}{76.8}$$
$$+ \frac{(2 - 6.4)^2}{6.4} = 28.776$$

$H_0$: good fit to binomial distribution with $\pi = .2$

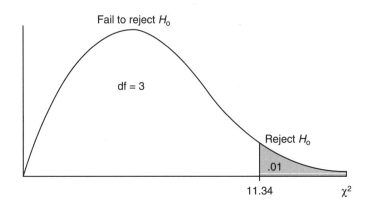

With df $= 4 - 1 = 3$ and $\alpha = .01$, the critical $\chi^2$-value is 11.34. Since $28.776 > 11.34$, there *is* sufficient evidence to reject $H_0$ and to conclude that the number of ads seen by each consumer does *not* follow a binomial distribution with $\pi = .2$.

**Example 19.6** Four commercial flights per day are made from a small county airport. The airport manager tabulates the number of on-time departures each day for 200 days.

Number of On-Time Departures

| | 0 | 1 | 2 | 3 | 4 |
|---|---|---|---|---|---|
| Observed number of days | 13 | 36 | 72 | 56 | 23 |

At the 5% significance level, test the null hypothesis that the daily distribution is binomial.

***Answer:*** We are not given a value of $\pi$ to test against, so we use the observed sample to find a proportion $\bar{p}$. The number of on-time departures is seen to be $13(0) + 36(1) + 72(2) + 56(3) + 23(4) = 440$. Since there were $4(200) = 800$ flights, we have a proportion of

$$\bar{p} = \frac{440}{800} = .55$$

The binomial distribution with $\pi = .55$ and $n = 4$ is

$$P(0) = (.45)^4 \qquad = .0410$$

$$P(1) = 4(.55)(.45)^3 \quad = .2005$$

$$P(2) = 6(.55)^2(.45)^2 = .3675$$

$$P(3) = 4(.55)^3(.45) \quad = .2995$$

$$P(4) = (.55)^4 \qquad = .0915$$

Multiplying each of these probabilities by 200 days gives the expected numbers of occurrences: $.0410(200) = 8.2$, $.2005(200) = 40.1$, $.3675(200) = 73.5$, $.2995(200) = 59.9$, $.0915(200) = 18.3$

Number of On-Time Departures

| | 0 | 1 | 2 | 3 | 4 |
|---|---|---|---|---|---|
| Expected number of days | 8.2 | 40.1 | 73.5 | 59.9 | 18.3 |

We calculate chi-square:

$$\chi^2 = \frac{(13 - 8.2)^2}{8.2} + \frac{(36 - 40.1)^2}{40.1} + \frac{(72 - 73.5)^2}{73.5} + \frac{(56 - 59.9)^2}{59.9}$$
$$+ \frac{(23 - 18.3)^2}{18.3} = 4.721$$

$H_o$: good fit with a binomial

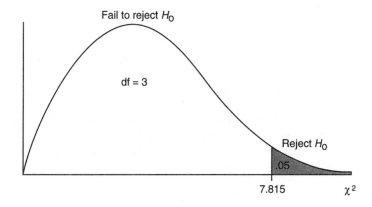

In this case the sample gives not only the total number, 200, but also the proportion, .55. Thus the number of degrees of freedom is the number of classes minus 2: $df = 5 - 2 = 3$. With $\alpha = .05$, we have a critical $\chi^2$-value of 7.815. Since $4.721 < 7.815$, there is *not* sufficient evidence to reject $H_o$. The observed data do *not* differ significantly from what would be expected for a binomial distribution.

# GOODNESS-OF-FIT TEST FOR POISSON DISTRIBUTION

**Example 19.7**  In a hospital study of 450 patients with a particular form of cancer, the following data shows the number of these patients who survived for each given number of years.

Survival Years

|  | 0 | 1 | 2 | 3 | 4 or more |
|---|---|---|---|---|---|
| Observed number of patients | 0 | 110 | 125 | 88 | 67 |

Test the null hypothesis that the distribution follows a Poisson distribution with $\mu = 2.1$. Assume a 2.5% significance level.

***Answer:*** The Poisson probabilities are

$$P(0) = e^{-2.1} \qquad = .122$$

$$P(1) = 2.1e^{-2.1} \qquad = .257$$

$$P(2) = \frac{(2.1)^2}{2} e^{-2.1} = .270$$

$$P(3) = \frac{(2.1)^3}{3!} e^{-2.1} = .189$$

$$P(4 \text{ or more}) = 1 - (.122 + .257 + .270 + .189) = .162$$

Multiplying each of these probabilities by 450 gives the number of patients expected to survive each designated number of years: $.122(450) = 54.9$, $.257(450) = 115.65$, $.270(450) = 121.5$, $.189(450) = 85.05$, $.162(450) = 72.9$.

Survival Years

|  | 0 | 1 | 2 | 3 | 4 or more |
|---|---|---|---|---|---|
| Expected number of patients | 54.9 | 115.65 | 121.5 | 85.05 | 72.9 |

Thus

$$\chi^2 = \frac{(60 - 54.9)^2}{54.9} + \frac{(110 - 115.65)^2}{115.65} + \frac{(125 - 121.5)^2}{121.5}$$
$$+ \frac{(88 - 85.05)^2}{85.05} + \frac{(67 - 72.9)^2}{72.9} = 1.430$$

$H_o$: good fit to Poisson distribution with $\mu = 2.1$

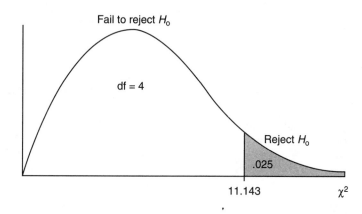

The number of degrees of freedom is the number of classes minus 1: $df = 5 - 1 = 4$. With $\alpha = .025$, the critical $\chi^2$-value is 11.143. Since $1.430 < 11.143$, there is *not* sufficient evidence to reject $H_o$. It should *not* be disputed that a Poisson with $\mu = 2.1$ describes the distribution of survival years of patients stricken with this cancer.

**Example 19.8** An editor checks the number of typing errors per page on 125 pages of a manuscript with the following results:

Number of Errors

|  | 0 | 1 | 2 | 3 | 4 | 5 |
|---|---|---|---|---|---|---|
| Observed<br># of pages | 29 | 38 | 24 | 18 | 13 | 3 |

Do these data follow a Poisson distribution? Test at the 10% significance level.

***Answer:*** In this case the mean is not given, so it must be estimated from the sample. The total number of errors is $29(0) + 38(1) + 24(2) + 18(3) + 10(4) + 3(5) = 195$. Thus the average per page of sample is 195/125 = 1.56. The Poisson probabilities are as follows:

$$P(0) = e^{-1.56} \qquad = .210$$

$$P(1) = 1.56e^{-1.56} \qquad = .328$$

$$P(2) = \frac{(1.56)^2}{2} e^{-1.56} = .256$$

$$P(3) = \frac{(1.56)^3}{3!} e^{-1.56} = .133$$

$$P(4) = \frac{(1.56)^4}{4!} e^{-1.56} = .052$$

$$P(5 \text{ or more}) = 1 - (.210 + .328 + .256 + .133 + .052) = .021$$

Multiplying by 125 gives the expected number of pages with various numbers of mistakes: .210(125) = 26.25, .328(125) = 41, .256(125) = 32, .133(125) = 16.625, .052(125) = 6.5, .021(125) = 2.625.[1]

Number of Errors

| | 0 | 1 | 2 | 3 | 4 | 5 |
|---|---|---|---|---|---|---|
| Expected # of pages | 26.25 | 41 | 32 | 16.625 | 6.5 | 2.625 |

$$\text{Thus } \chi^2 = \frac{(29 - 26.25)^2}{26.25} + \frac{(38 - 41)^2}{41} + \frac{(24 - 32)^2}{32}$$
$$+ \frac{(18 - 16.625)^2}{16.625} + \frac{(13 - 6.5)^2}{6.5} + \frac{(3 - 2.625)^2}{2.625}$$
$$= 9.175$$

$H_o$: good fit with a Poisson distribution

Since the sample is giving both the total number of pages (125) and the average errors per page (1.56), the number of degrees of freedom is the number of classes minus 2: $df = 6 - 2 = 4$. With $\alpha = .10$ the critical $\chi^2$-value is 7.779. Since 9.175 > 7.779, there *is* sufficient evidence at the 10% level to reject $H_o$ and to claim that the typing errors per page do *not* follow a Poisson distribution.

---

[1]Note this does violate the previously mentioned rule of thumb that every cell must have an expected value of at least 5. Some statisticians accept one lower value; alternatively, we could combine the last two cells into one cell with a higher expected value.

# GOODNESS-OF-FIT TEST FOR NORMAL DISTRIBUTION

**Example 19.9** A purchasing agent weighs a sample of 225 bags of rice labeled as containing 50 pounds each with the following results.

|  | Weight (Pounds) | | | | |
|---|---|---|---|---|---|
|  | Under 49.25 | 49.25 −49.75 | 49.75 −50.25 | 50.25 −50.75 | Over 50.75 |
| Observed number of bags | 25 | 61 | 70 | 59 | 10 |

Test the null hypothesis that the data follow a normal distribution with $\mu$ = 50 and $\sigma$ = 0.5. Assume a 2.5% and then a 1% significance level.

**Answer:** The z-scores for 50.25 and 50.75 are, respectively,

$$\frac{50.25 - 50}{0.5} = 0.5 \quad \text{and} \quad \frac{50.75 - 50}{0.5} = 1.5$$

Similarly, 49.75 and 49.25 have z-scores of −0.5 and −1.5. Using Table A, we find these values:

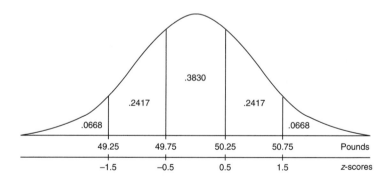

Multiplying each probability by 225 gives the expected numbers of bags for the corresponding weight ranges:

.0668(225) = 15.03, .2417(225) = 54.38, .3830(225) = 86.18.

Weight (pounds)

|  | Under 49.25 | 49.25 −49.75 | 49.75 −50.25 | 50.25 −50.75 | Over 50.75 |
|---|---|---|---|---|---|
| Expected number of bags | 15.03 | 54.38 | 86.18 | 54.38 | 15.03 |

Thus $\chi^2 = \dfrac{(25 - 15.03)^2}{15.03} + \dfrac{(61 - 54.38)^2}{54.38} + \dfrac{(70 - 86.18)^2}{86.18}$
$+ \dfrac{(59 - 54.38)^2}{54.38} + \dfrac{(10 - 15.03)^2}{15.03} = 12.533$

$H_0$: good fit to a normal distribution with $\mu = 50$ and $\sigma = 0.5$

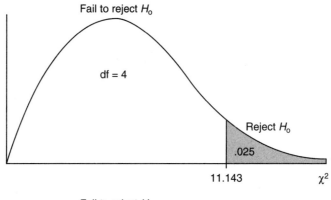

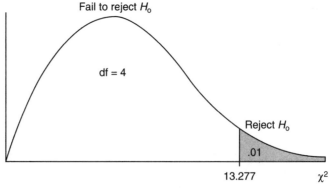

With $df = 5 - 1 = 4$ and $\alpha = .025$, the critical $\chi^2$-value is 11.143, while $\alpha = .01$ results in a critical value of 13.277. Since 12.533 > 11.143 and 12.533 < 13.277, we have sufficient evidence to reject $H_0$ at the 2.5% significance level but *not* at the 1% level. Thus, if the purchasing agent is willing to accept a 2.5% $\alpha$-risk, she should conclude that the distribution of weights of rice bags is *not* normal.

**Example 19.10** Suppose that the assembly times for a sample of 300 units of an electronic product have mean $\mu = 84$, standard deviation $\sigma = 3$, and the following distribution.

Assembly Time (minutes)

|  | <78 | 78-81 | 81-84 | 84-87 | 87-90 | >90 |
|---|---|---|---|---|---|---|
| Observed number of units | 15 | 39 | 87 | 96 | 48 | 15 |

At the 1% significance level, test the null hypothesis that the distribution is normal.

***Answer:*** Here we must use the sample mean, 84, and the sample standard deviation, 3. Table A gives these values:

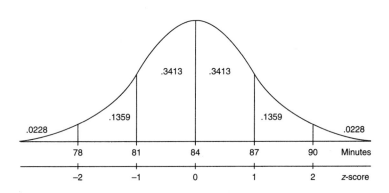

Multiplying by 300 units yields the following expected numbers of units:

Assembly Time (minutes)

|  | <78 | 78-81 | 81-84 | 84-87 | 87-90 | >90 |
|---|---|---|---|---|---|---|
| Expected number of units | 6.84 | 40.77 | 102.39 | 102.39 | 40.77 | 6.84 |

$$\text{Thus } \chi^2 = \frac{(15 - 6.84)^2}{6.84} + \frac{(39 - 40.77)^2}{40.77} + \frac{(87 - 102.39)^2}{102.39}$$
$$+ \frac{(96 - 102.39)^2}{102.39} + \frac{(48 - 40.77)^2}{40.77} + \frac{(15 - 6.84)^2}{6.84}$$
$$= 23.540$$

Since we are using three measures from the sample (size, mean, and standard deviation), the number of degrees of freedom equals the number of classes minus 3: $df = 6 - 3 = 3$. With $\alpha = .01$, the critical $\chi^2$-value is 11.34. Since $23.540 > 11.34$, there *is* sufficient evidence to reject $H_o$ and to conclude that the distribution of assembly times is *not* normal.

# FITS THAT ARE TOO GOOD

We have been using only small, upper-tail probabilities of the chi-square distribution. In some cases, however, we are interested in the probability that a chi-square value is less than or equal to a critical value.

**Example 19.11** Suppose your mathematics instructor gives, as a homework assignment, the problem of testing the fairness of a certain die. You are asked to roll the die 6000 times and to note how often it comes up 1, 2, 3, 4, 5, and 6. Tossing the die soon becomes tiresome, so you decide to invent some "reasonable" data. Being careful to make the total 6000, you write:

Number on Die

| | 1 | 2 | 3 | 4 | 5 | 6 |
|---|---|---|---|---|---|---|
| Observed | 988 | 991 | 1010 | 990 | 1013 | 1008 |

$H_o$: good fit to uniform distribution; that is, die is fair

The "expected" values are all 1000, so

$$\chi^2 = \frac{12^2}{1000} + \frac{9^2}{1000} + \frac{10^2}{1000} + \frac{10^2}{1000} + \frac{13^2}{1000} + \frac{8^2}{1000} = 0.658$$

a. What is the conclusion if $\alpha = .01$? .05? .10?

**Answer:** With $df = 6 - 1 = 5$, the critical $\chi^2$-values are 15.086, 11.070, and 9.236. Since 0.658 is less than each of these, there is not enough evidence to claim that the die is unfair.

b. What is the conclusion if we're willing to accept $\alpha = .90$? .95? .975? .99?

**Answer:** These give critical $\chi^2$-values of 1.610, 1.145, 0.831, and 0.554, respectively, so only if we accept a 99% chance of committing a Type I error can we reject $H_0$ and question the fairness of the die. Even if the die is fair, the probability of obtaining such a good-fitting sample is extremely small. Your instructor would reasonably conclude that you simply made up the data. Your results are too good to be true!

Statisticians have shown that some of Gregor Mendel's experimental results are too good to be true. No one disputes his conclusions, but there are strong indications that the monks working for him falsified some data to conform with the intended conclusions.

# EXERCISES

1. A college alumni office states that 20% of the graduates eventually become lawyers; 25%, doctors; and 35%, corporate executives. The remaining 20% are spread among a variety of professions. A new survey taken of 625 graduates turned up 110 lawyers, 140 doctors, 250 corporate executives, and 125 others. Is there sufficient evidence that the percentages quoted by the alumni office have changed? Test at the 5% significance level.

2. A highway superintendent states that five bridges into a city are used in the ratio 2:3:3:4:6 during the morning rush hour. A highway study of a sample 6000 cars indicates that 720, 970, 1013, 1380, and 1917 cars use the five bridges, respectively. Can the superintendent's claim be rejected if $\alpha = .05$? .025?

3. A study (*The Lancet*, October 22, 1994, page 1137) of accident records at a large engineering company in England reported the following number of injuries on each shift for 1 year:

| | Shift | | |
|---|---|---|---|
| | **Morning** | **Afternoon** | **Night** |
| Number of injuries | 1372 | 1578 | 1686 |

Is there sufficient evidence to say that the numbers of accidents on the three shifts are not the same? Test with $\alpha = .05$.

4. A building inspector inspects five major construction sites per day. In a sample 200 days, the numbers of sites passing inspection are as follows:

Number of Sites Passing

| | ≤2 | 3 | 4 | 5 |
|---|---|---|---|---|
| Frequency | 5 | 34 | 90 | 71 |

At a 5% significance level can we conclude that the distribution is described by a binomial with $\pi = .8$?

5. A sampling of 500 major corporations is examined to determine the number of women among the top three executives. Using $\alpha = .01$, test the null hypothesis that the results tabulated below follow a binomial distribution. Note that you must use the observed data to approximate the population proportion $\pi$.

Number of Women
Among Top 3 Executives

| | 0 | 1 | 2 | 3 |
|---|---|---|---|---|
| Number of corporations | 245 | 170 | 50 | 35 |

6. A real estate agent tabulates the number of home sales per week during a sample of 83 weeks.

Sales per Week

| | 0 | 1 | 2 | 3 | 4 | 5 or more |
|---|---|---|---|---|---|---|
| Frequency | 3 | 15 | 28 | 25 | 6 | 4 |

Test the null hypothesis that the distribution is Poisson with $\mu = 2.4$. Use $\alpha = .05$.

7. A company efficiency expert records the number of wrong connections per hour by a new telephone switchboard operator. For a sample of 50 hours, the results are as follows:

Mistakes per Hour

| | 0 | 1 | 2 | 3 | 4 |
|---|---|---|---|---|---|
| Frequency | 5 | 16 | 17 | 8 | 4 |

Test the null hypothesis that the distribution is Poisson. Assume a 10% significance level, and note that you must use the observed data to approximate $\mu$ .

8. Consider the following distribution of salaries of 100 junior level executives:

Salary ($1000)

| | <40 | 40-50 | 50-60 | 60-70 | >70 |
|---|---|---|---|---|---|
| Frequency | 10 | 32 | 33 | 15 | 10 |

Test the null hypothesis that the distribution is normal with a mean of 55 and a standard deviation of 10. Use $\alpha = .025$.

## Study Questions

1. Consider three of our often used $z$-scores: 1.645, 1.96, and 2.576. Square these values, and look for the results in Table D, the chi-square distribution table, in the Appendix. Conclusion?

2. Show that

$$\Sigma \frac{(\text{obs} - \text{exp})^2}{\text{exp}} = \Sigma \frac{(\text{obs})^2}{\text{exp}} - N$$

where $N$ is the total number of elements in the sample. [Hint: $\Sigma$ exp $= \Sigma$ obs $= N]$

# 20
# CHI-SQUARE TEST FOR INDEPENDENCE

In each of the goodness-of-fit problems in Chapter 19, a set of expectations was based on some assumption about how the distribution should turn out. We then tested whether an observed sample distribution might reasonably have come from a larger set based on the assumed distribution.

In many real-world problems, however, we want to compare two or more observed samples without any prior assumptions about an expected distribution. In what is called a *test of independence*, we ask whether the two or more samples might reasonably have come from some one larger set. For example, do students, professors, and administrators all have the same opinion concerning the need for a new science building? Do nonsmokers, light smokers, and heavy smokers all have the same likelihood of being eventually diagnosed with cancer, heart disease, or emphysema?

We classify our observations in two ways, and then ask whether the two ways are independent of each other. For example, we might consider several age groups, and within each group ask how many employees show various levels of job satisfaction. The null hypothesis would be that age and job satisfaction are independent; that is, that the proportion of employees expressing a given level of job satisfaction is the same no matter which age group is considered. A sociologist might classify people by ethnic origin, and within each group ask how many individuals complete various levels of education. The null hypothesis would be that ethnic origin and education level are independent; that is, that the proportion of people achieving a given education level is the same no matter which ethnic group is considered.

Our analysis will involve calculating a table of *expected* values, assuming the null hypothesis about independence is true. We then compare these expected values with the observed values, and ask whether the differences are reasonable if $H_0$ is true. The significance of the differences is gauged by the same $\chi^2$-value of weighted squared differences used in Chapter 19. The smaller the resulting $\chi^2$-value, the more reasonable is the null hypothesis of independence. If the $\chi^2$-value exceeds some critical value, we can say that the evidence is sufficient to reject the null hypothesis and to claim that there *is* some relationship between the two variables or methods of classification.

**Example 20.1** A beef distributor wishes to determine whether there is a relationship between geographic region and cut of meat preferred. If there is no relationship, we will say that beef preference is *independent* of geographic region. Suppose that in a random sample of 500 consumers, 300 are from the North while 200 are from the South. Of these 500 persons, 150 prefer cut A, 275 prefer cut B, and only 75 prefer cut C.

Geographic Region

| Beef preference | North | South | |
|---|---|---|---|
| Cut A | | | 150 |
| Cut B | | | 275 |
| Cut C | | | 75 |
| | 300 | 200 | |

a. If beef preference is independent of geographic region, how would we expect this table to be filled in?

**Answer:** Since 300/500 = .6 of the sample is from the North, we would expect .6 of the 150 consumers favoring cut A to be from the North. We calculate .6(150) = 90 or (300)(150)/500 = 90. Similarly, we would expect .6 of the 275 consumers favoring cut B to be from the North. We calculate .6(275) = 165 or (300)(275)/500 = 165. Continuing in this manner we fill in the table as follows:

| Expected results: | North | South | |
|---|---|---|---|
| Cut A | 90 | 60 | 150 |
| Cut B | 165 | 110 | 275 |
| Cut C | 45 | 30 | 75 |
| | 300 | 200 | |

Note that we didn't have to perform all the calculations to fill in the table. After determining 90 and 165, no choices were left for the remaining values because the row totals and column totals were already set. Thus, in this problem there are *two* degrees of freedom, that is, $df = 2$.

b. Now suppose that in the actual sample of 500 consumers, the observed numbers were as follows:

|  |  | North | South |  |
|---|---|---|---|---|
|  | Cut $A$ | 100 | 50 | 150 |
| Observed results: | Cut $B$ | 150 | 125 | 275 |
|  | Cut $C$ | 50 | 25 | 75 |
|  |  | 300 | 200 |  |

Are the differences between the expected and observed values large or small? If the differences are large enough, we will reject independence and claim that beef preference is related to geographic location.

***Answer:*** We calculate the $\chi^2$ value:

$$\chi^2 = \Sigma \frac{(\text{obs} - \text{exp})^2}{\text{exp}}$$

$$= \frac{(100-90)^2}{90} + \frac{(50-60)^2}{60} + \frac{(150-165)^2}{165} + \frac{(125-110)^2}{110}$$

$$+ \frac{(50-45)^2}{45} + \frac{(25-30)^2}{30} = 7.578$$

Is 7.578 large enough for us to reject the null hypothesis of independence?

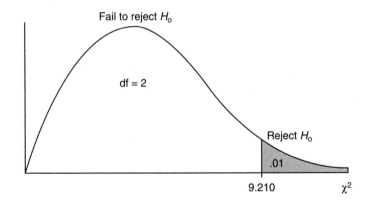

Looking at the $\chi^2$ table, with $df = 2$, we see that, with $\alpha = .01$, the critical $\chi^2$-value is 9.210. Since $7.578 < 9.210$, there is *not* enough evidence to reject $H_o$.

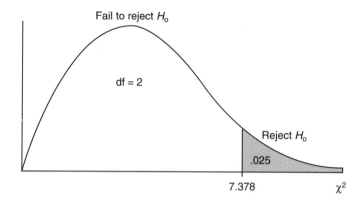

However, with $\alpha = .025$, the critical $\chi^2$-value is 7.378; and since $7.578 > 7.378$, there *is* sufficient evidence to reject $H_o$. Thus, we can conclude that at the 2.5% significance level there *is* a relationship between beef preference and geographic region, but at the 1% significance level the evidence is *not* sufficient to reject the null hypothesis of independence.

# CONTINGENCY TABLES

**Example 20.2**   In a nationwide telephone poll of 1000 adults, representing Democrats, Republicans, and Independents, respondents were asked if their confidence in the U.S. banking system had been shaken by the Savings and Loan crisis. The answers, cross-classified by party affiliation, are given in the following *contingency* table.

|  | Observed: | | |
|---|---|---|---|
|  | Yes | No | No opinion |
| Democrats | 175 | 220 | 55 |
| Republicans | 150 | 165 | 35 |
| Independents | 75 | 105 | 20 |

Test the null hypothesis that shaken confidence in the banking system is independent of party affiliation. Use a 10% significance level.

***Answer:*** The above table gives the *observed* results. To determine the *expected* values, we must first determine the row and column totals, which were given in the Example 20.1. We obtain the following values:

Row totals:  $175 + 220 + 55 = 450, 150 + 165 + 35 = 350,$
 $75 + 105 + 20 = 200$
Column totals:  $175 + 150 + 75 = 400, 220 + 165 + 105 = 490,$
 $55 + 35 + 20 = 110$

|  | Yes | No | No opinion |  |
|---|---|---|---|---|
| Democrats |  |  |  | 450 |
| Republicans |  |  |  | 350 |
| Independents |  |  |  | 200 |
|  | 400 | 490 | 110 |  |

To calculate, for example, the expected value in the upper left box, we can proceed in any of several equivalent ways. First, we could note that the proportion of Democrats is $450/1000 = .45$; and so, if independent, the expected number of Democrat *yes* responses is $.45(400) = 180$. Instead, we could note that the proportion of *yes* responses is $400/1000 = .4$; and so, if independent, the expected number of Democrat *yes* responses is $.4(450) = 180$. Finally, we could note that both these calculations simply involve $(450)(400)/1000 = 180$.

In other words, *the expected value of any box can be calculated by multiplying the corresponding row total times the appropriate column total and then dividing by the grand total.* Thus, for example, the expected value for the middle box, which corresponds to Republican *no* responses, is $(350)(490)/1000 = 171.5$.

Continuing in this manner, we fill in the table as follows:

|  |  | Expected: |  |  |
|---|---|---|---|---|
|  | Yes | No | No opinion |  |
| Democrats | 180 | 220.5 | 49.5 | 450 |
| Republicans | 140 | 171.5 | 38.5 | 350 |
| Independents | 80 | 98 | 22 | 200 |
|  | 400 | 490 | 110 |  |

$$\text{Then } \chi^2 = \frac{(175 - 180)^2}{180} + \frac{(220 - 220.5)^2}{220.5} + \frac{(55 - 49.5)^2}{49.5}$$
$$+ \frac{(150 - 140)^2}{140} + \frac{(165 - 171.5)^2}{171.5} + \frac{(35 - 38.5)^2}{38.5}$$
$$+ \frac{(75 - 80)^2}{80} + \frac{(100 - 98)^2}{98} + \frac{(20 - 22)^2}{22}$$
$$= 3.024$$

Note that, once the 180, 220.5, 140, and 171.5 boxes are calculated, the other expected values can be found by using the row and column totals. Thus, the number of degrees of freedom here is 4. More generally, in this type of problem

$$df = (r - 1)(c - 1)$$

where $r$ is the number of rows and $c$ is the number of columns.

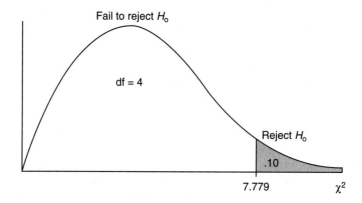

With $\alpha = .10$ and $df = 4$, the critical $\chi^2$-value is 7.779. Since 3.024 $< 7.779$, there is *not* sufficient evidence to reject the null hypothesis of independence. Thus, at the 10% significance level, we *cannot* claim a relationship between party affiliation and shaken confidence in the banking system.

Two points are worth noting:

First, if any of the *expected* values is too small, the $\chi^2$-value tends to turn out to be unfairly high. The usual rule-of-thumb cutoff point is taken as 5; that is, this procedure is not used if any expected value is below 5. Sometimes this difficulty is overcome by a regrouping that combines two or more classifications. However, such combination leads to smaller numbers of degrees of freedom and less information from the tables.

Second, even if there is sufficient evidence to reject the null hypothesis of independence, we cannot necessarily claim any direct *causal* relationship. In other words, although we can make a statement about some link or relationship between two variables, we are *not* justified in claiming that one *causes* the other. For example, we may demonstrate a relationship between salary level and job satisfaction, but our methods would not show that higher salaries *cause* higher job satisfaction. Perhaps an employee's higher job satisfaction impresses his superiors and thus leads to larger increases in pay. Or perhaps there is a third variable, such as training, education, or personality, that has a direct causal relationship with both salary level and job satisfaction.

Example 20.3 illustrates the step-by-step analysis of a contingency table.

**Example 20.3** A medical researcher tests 640 heart-attack victims for the presence of a certain antibody in their blood and cross-classifies against the severity of the attack. The results are reported in the following table:

|  |  | Severity of Attack | | |
|---|---|---|---|---|
|  |  | Severe | Medium | Mild |
| Antibody test | Positive | 85 | 125 | 150 |
|  | Negative | 40 | 95 | 145 |

Is there evidence of a relationship between presence of the antibody and severity of the heart attack? Test at the 5% significance level.

**Answer:**

$H_0$: independence (no relation between presence of antibody and severity of attack)

$H_a$: dependence (severity of attack related to presence of antibody)

$$\alpha = .05$$

$$df = (2 - 1)(3 - 1) = 2$$

Totaling rows and columns yields these results:

|  | Severe | Medium | Mild |  |
|---|---|---|---|---|
| Positive |  |  |  | 360 |
| Negative |  |  |  | 280 |
|  | 125 | 220 | 295 |  |

The expected value for each box is calculated by multiplying the row total by the column total and dividing by 640:

(360)(125)=70.3,   (360)(220)=123.8,   (360)(295)=165.9,
(280)(125)=54.7,   (280)(220)=96.2,   (280)(295)=129.1

|  | Severe | Expected: Medium | Mild |  |
|---|---|---|---|---|
| Positive | 70.3 | 123.8 | 165.9 | 360 |
| Negative | 54.7 | 96.2 | 129.1 | 280 |
|  | 125 | 220 | 295 |  |

Then $\chi^2 = \dfrac{(85-70.3)^2}{70.3} + \dfrac{(125-123.8)^2}{123.8} + \dfrac{(150-165.9)^2}{165.9}$
$+ \dfrac{(40-54.7)^2}{54.7} + \dfrac{(95-96.2)^2}{96.2} + \dfrac{(145-129.1)^2}{129.1}$
$= 10.533$

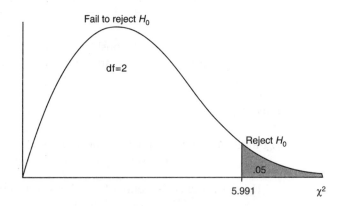

With $\alpha = .05$ and df = 2, the critical $\chi^2$-value is 5.991. Since 10.533 > 5.991, there *is* sufficient evidence to reject the null hypothesis of independence. Thus, at the 5% significance level, there *is* a relationship between presence of the antibody and severity of the heart attack.

# EXERCISES

1. A surprising study of 1437 male hospital admissions reported in *The New York Times* (February 24, 1993, page C12) found that, of 665 patients admitted with heart attacks, 214 had vertex baldness, while of the remaining

772 non-heart-related admissions, 175 had vertex baldness. Is this evidence sufficient at the 5% significance level to say that there is a relationship between heart attacks and vertex baldness?

2. Following are the results of a study (*New England Journal of Medicine*, 1988) to determine whether taking an aspirin every other day reduces the risk of heart attacks in men:

|                 | Aspirin | Placebo |
|-----------------|---------|---------|
| Fatal attack    | 5       | 18      |
| Nonfatal attack | 99      | 171     |
| No heart attack | 10,933  | 10,845  |

Is there evidence of a relationship between taking aspirins and the risk of heart attacks? Test at the 5% significance level.

3. A 4 year study, reported in *The New York Times*, of men more than 70 years old analyzed blood cholesterol, and noted how many men with different cholesterol levels suffered nonfatal or fatal heart attacks.

|                        | Low Cholesterol | Medium Cholesterol | High Cholesterol |
|------------------------|-----------------|--------------------|------------------|
| Nonfatal heart attacks | 29              | 17                 | 18               |
| Fatal heart attacks    | 19              | 20                 | 9                |

Is there evidence of a relationship? Does level of cholesterol seem to predict risk of heart disease for men over 70 years old? Test at the 10% significance level.

4. Does sea bathing lead to illness? For a sampling of 1112 bathers in four English resort areas (*The Lancet*, October 1, 1994, page 906), the following table shows exposure to varying amounts of fecal material in the water versus number of cases of gastroenteritis reported.

|                           | Fecal Streptococci per 100 ml | | | |
|---------------------------|------|-------|-------|---------|
|                           | None | 0–34  | 35–69 | over 70 |
| Cases of gastroenteritis  | 57   | 34    | 24    | 17      |
| No related illness        | 548  | 273   | 125   | 34      |

Is there evidence of a relationship between amount of fecal material in the water and reported gastroenteritis illness? Test at the 1% significance level.

5. Is hair-loss pattern related to body mass index (kg/m²)? One study (*Journal of the American Medical Association*, February 24, 1993, page 1000) of 769 men showed the following numbers:

|  |  | Hair-Loss Pattern | | |
|---|---|---|---|---|
|  |  | None | Frontal | Vertex |
| Body-mass index | <25 | 137 | 22 | 40 |
|  | 25−28 | 218 | 34 | 67 |
|  | >28 | 153 | 30 | 68 |

At the 5% significance level is there a relationship between hair-loss pattern and body-mass index?

6. Is there a relationship between fitness level and smoking habits? A study (*New England Journal of Medicine*, February 25, 1993, page 535) cross-classified individuals in four categories of fitness and four categories of smoking habits:

|  | Fitness Level | | | |
|---|---|---|---|---|
|  | Low | Medium | | High |
| Never smoked | 113 | 113 | 110 | 159 |
| Former smokers | 119 | 135 | 172 | 190 |
| 1 to 9 cigarettes daily | 77 | 91 | 86 | 65 |
| ≥10 cigarettes daily | 181 | 152 | 124 | 73 |

Is there a relationship between fitness level and smoking habits? Test at the 2.5% significance level.

## Study Questions

1. Suppose that in a telephone survey of 800 registered voters, the data are cross-classified both by sex of respondent and by respondent's opinion on an environmental bond issue.

|  | Bond Issue | |
|---|---|---|
|  | For | Against |
| Men | 450 | 150 |
| Women | 160 | 40 |

a. Determine the $\chi^2$ value for these data.
b. Suppose that instead we use the approach given in Chapter 18 to test the significance of the difference between the proportions
$\bar{p}_1 = 450/600 = .75$ and $\bar{p}_2 = 160/200 = .8$.
Let $\bar{p} = (450+160)/800 = .7625$ and calculate

$$\sigma_d = \sqrt{(.7625)(.2375)\left(\frac{1}{600} + \frac{1}{200}\right)}$$

c. Find the observed $z$-value which equals $(.75 - .8)/\sigma_d$.
d. Show that the square of this $z$-value is precisely equal to the $\chi^2$ value determined in part a!
It can be shown more generally that the $\chi^2$ value for a two-by-two table is equal to the squared observed $z$-value for the difference in proportions. In this case we can determine the $p$-value for a $\chi^2$ value by taking the square root of the $\chi^2$ value and then using the normal distribution table.

2. A survey of college students focuses on a cross-classification by year in school and positive or negative feelings about the college experience.

| | Year in School | | | |
| --- | --- | --- | --- | --- |
| | Freshman | Sophomore | Junior | Senior |
| Positive | 105 | 100 | 90 | 125 |
| Negative | 35 | 50 | 30 | 25 |

a. Determine the value of $\chi^2$.
b. There are three degrees of freedom in this example. Suppose the table is split into the following three smaller tables, each having a single degree of freedom:

| | Fresh. | Soph. | | Jun. | Sen. | | Undercl. | Uppercl. |
| --- | --- | --- | --- | --- | --- | --- | --- | --- |
| Pos. | 105 | 100 | Pos. | 90 | 125 | Pos. | 205 | 215 |
| Neg. | 35 | 50 | Neg. | 30 | 25 | Neg. | 85 | 55 |

Determine the three $\chi^2$ values, one from each of these tables.
c. Show that the $\chi^2$ value of the original table is equal to the sum of the three $\chi^2$ values calculated. This illustrates the *additive property* of the $\chi^2$ statistic.

# Part 7

# REGRESSION

*Many decisions are based on a perceived relationship between two variables. For example, a company's market share may vary directly with advertising expenditures. A person's blood pressure may increase or decrease inversely as less or more hypertension medication is taken. A student's grades will probably go up and down according to the number of hours of studying time per week.*

*Two questions arise. First, how can the strength of an apparent relationship be measured? Second, how can an observed relationship be put into functional terms? For example, a real estate broker not only might wish to determine whether a relationship exists between the prime rate and the number of new homes sold in a month, but also would find useful an expression with which to predict the number of house sales given a particular value of the prime rate.*

*Chapters 21 and 22 deal with the related topics of* scatter diagrams, linear regression, predictions using the regression line, correlation coefficients, *and* hypothesis tests for correlation.

# 21
# LINEAR REGRESSION

Our studies so far have been concerned with measurements of a single variable such as the mean or proportion. However, many important applications of statistics involve examining whether two or more variables are related to one another. For example, is there a relationship between the smoking histories of pregnant women and the birth weights of their children? Between SAT scores and success in college? Between amount of fertilizer used and amount of crop harvested? Between price and numbers of bedrooms and bathrooms in new homes?

## SCATTER DIAGRAMS

Suppose a relationship is perceived between two variables $X$ and $Y$, and we graph the pairs $(x,y)$. The result, called a *scatter diagram*, gives a visual impression of the existing relationship between the variables. In this chapter we ask whether the relationship can be reasonably explained in terms of a linear function, that is, one whose graph is a straight line.

For example, we might be looking at a picture such as this:

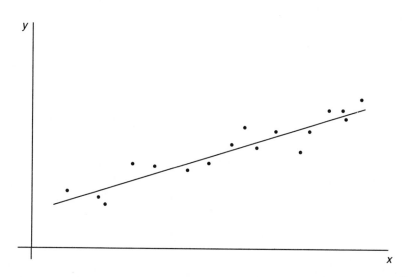

We need to know what the term *best-fitting straight line* means and how we can find this line. Furthermore, we want to be able to gauge whether the relationship between the variables is strong enough so that finding and making use of this straight line is meaningful.

We will use $(x,y)$ to represent observed data points and $(x,y')$ to signify the corresponding points on the *best-fitting* straight line. For example:

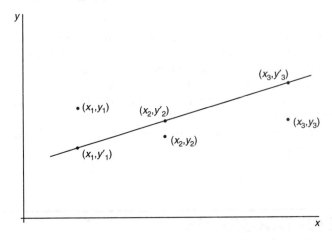

It is reasonable, intuitive, and correct that the best-fitting line will pass through $(\bar{x}, \bar{y})$ where $\bar{x}$ and $\bar{y}$ are the means of the variables $X$ and $Y$. Then, from the basic expression for a line with a given slope through a given point, we have the equation

$$y' - y = m(x - \bar{x})$$

or    $$y' = \bar{y} + m(x - \bar{x})$$

where $m$ is the slope of the line.

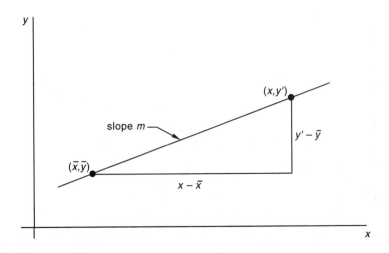

**Example 21.1**   The ages and salaries (in thousands of dollars) of four executives in a business firm are as follows:

| Age: | 38 | 53 | 42 | 47 |
|------|----|----|----|----|
| Salary: | 45 | 86 | 58 | 61 |

Plotting the four points (38,45), (53,86), (42,58), and (47,61) gives the following *scatter diagram*:

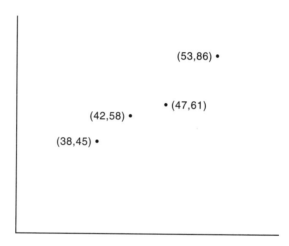

What is the best fitting straight line that can be drawn above?

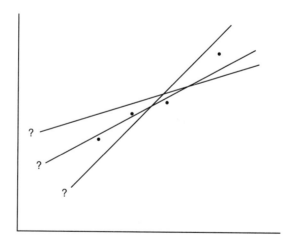

***Answer:*** On the basis of our experiences with measuring variances, by best fitting we will mean the straight line that minimizes the sum of the squares of the differences between the observed values and the values predicted by the line.

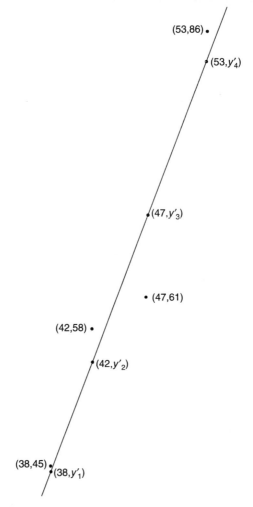

(53,86) •

(53,$y'_4$)

(47,$y'_3$)

• (47,61)

(42,58) •

(42,$y'_2$)

(38,45) •

(38,$y'_1$)

In other words, we want to minimize

$$(45 - y'_1)^2 + (58 - y'_2)^2 + (61 - y'_3)^2 + (86 - y'_4)^2$$

The mean of the $x$'s is $\bar{x} = (38 + 53 + 42 + 47)/4 = 45$ and the mean of the $y$'s is $\bar{y} = (45 + 86 + 58 + 61)/4 = 62.5$. The best fitting straight line will be of the form

$$y' = 62.5 + m(x - 45)$$

Thus

$$y'_1 = 62.5 + m(38 - 45) = 62.5 - 7m$$

$$y'_2 = 62.5 + m(53 - 45) = 62.5 + 8m$$

$$y'_3 = 62.5 + m(42 - 45) = 62.5 - 3m$$

$$y'_4 = 62.5 + m(47 - 45) = 62.5 + 2m$$

We want to minimize

$$(62.5 - 7m - 45)^2 + (62.5 + 8m - 86)^2$$
$$+ (62.5 - 3m - 58)^2 + (62.5 + 2m - 61)^2$$
$$= (17.5 - 7m)^2 + (-23.5 + 8m)^2 + (4.5 - 3m)^2 + (1.5 + 2m)^2$$
$$= 126m^2 - 642m + 881$$

Remembering that the parabola $am^2 + bm + c$ has its lowest point (its vertex) when $m = -b/2a$, we find that $126m^2 - 642m + 881$ is minimized when $m = -(-642/252) = 2.548$.

Thus the equation of the best fitting line is

$$y' = 62.5 + 2.548(x - 45)    \text{or}$$

$$y' = 2.548x - 52.16$$

So, for example, we might predict that the salary of a 50-year-old executive is approximately $2.548(50) - 52.6 = 74.8$ thousand dollars ($74,800).

It is also instructive to ask about the meaning of the slope $m$. In this example, 2.548 thousand dollars ($2548) is a representative change in salary for each year of increase in the age of an executive.

# EQUATION OF THE REGRESSION LINE

The best fitting straight line, that is, the line that minimizes the sum of the squares of the differences between the observed values and the values predicted by the line, is called the *regression line*. We generalize the procedure from Example 21.1.

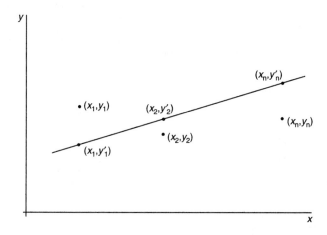

We wish to minimize

$$(y_1 - y'_1)^2 + (y_2 - y'_2)^2 + \ldots + (y_n - y'_n)^2$$

Using the expression $y_i = \bar{y} + m(x_i - \bar{x})$, from the preceding section, and, again, the fact that $am^2 + bm + c$ has its lowest point when $m = -b/2a$, we can show algebraically:

$$m = \frac{\sum xy - n\bar{x}\bar{y}}{\sum x^2 - n(\bar{x})^2} \text{ where}$$

$$\sum xy = x_1 y_1 + x_2 y_2 + \ldots + x_n y_n \quad \text{and} \quad \sum x^2 = x_1^2 + x_2^2 + \ldots + x_n^2$$

Then the equation of the regression line, as before, is

$$y' = \bar{y} + m(x - \bar{x})$$

# INTERPRETATION OF SLOPE AND PREDICTIONS USING REGRESSION LINE

**Example 21.2**   An insurance company conducts a survey of 15 of its life insurance agents. The average number of minutes spent with each potential customer and the number of policies sold in a week are noted for each agent. Letting $X$ and $Y$ represent the average number of minutes and the number of sales, respectively, we have:

| X: | 25 | 23 | 30 | 25 | 20 | 33 | 18 | 21 | 22 | 30 | 26 | 26 | 27 | 29 | 20 |
|----|----|----|----|----|----|----|----|----|----|----|----|----|----|----|----|
| Y: | 10 | 11 | 14 | 12 | 8 | 18 | 9 | 10 | 10 | 15 | 11 | 15 | 12 | 14 | 11 |

Find the equation of the best-fitting straight line for the data.

**Answer:** Plotting the 15 points (25,10), (23,11), . . . , (20,11) gives an intuitive visual impression of the relationship:

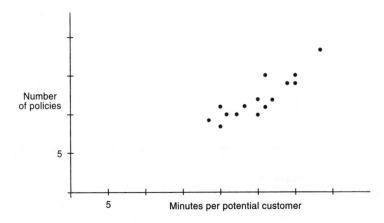

This scatter diagram indicates the existence of a relationship that appears to be *linear*; that is, the points lie roughly on a straight line. Furthermore, the linear relationship is *positive*; that is, as one variable increases, so does the other (the straight line slopes upward).

To determine the equation of the best-fitting line, we first calculate:

$$\Sigma x = 25 + 23 + 30 + \ldots + 20 = 375$$

$$\Sigma y = 10 + 11 + 14 + \ldots + 11 = 180$$

$$\Sigma x^2 = 25^2 + 23^2 + 30^2 + \ldots + 20^2 = 9639$$

$$\Sigma xy = 25(10) + 23(11) + 30(14) + \ldots + 20(11) = 4645$$

Then $n = 15$, so

$$\bar{x} = \frac{\Sigma x}{n} = \frac{375}{15} = 25, \qquad \bar{y} = \frac{\Sigma y}{n} = \frac{180}{15} = 12$$

and

$$m = \frac{\Sigma xy - n\bar{x}\bar{y}}{\Sigma x^2 - n(\bar{x})^2} = \frac{4645 - 15(25)(12)}{9639 - 15(25)^2} = \frac{145}{264} = 0.5492$$

The regression line is given by

$$y' = \bar{y} + m(x - \bar{x}) = 12 + 0.5492(x - 25) = 0.5492x - 1.73$$

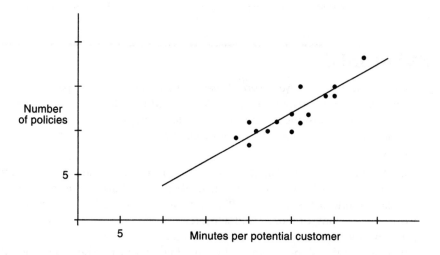

Number of policies

5

5    Minutes per potential customer

Thus, for example, we might predict that agents who average 24 minutes per customer will average $0.5492(24) - 1.73 = 11.45$ sales per week. We also note that each additional minute spent seems to produce an average 0.5492 extra sale.

**Example 21.3**   Following are advertising expenditures and total sales with regard to six detergent products:

| Advertising ($1000): x | 2.3 | 5.7 | 4.8 | 7.3 | 5.9 | 6.2 |
|---|---|---|---|---|---|---|
| Total sales ($1000): y | 77 | 105 | 96 | 118 | 102 | 95 |

a. Find the equation of the regression line.

**Answer:** We calculate $\Sigma x = 32.2$, $\Sigma y = 593$, $\Sigma xy = 3288.6$, and $\Sigma x^2 = 187.36$ which give $\bar{x} = 5.367$, $\bar{y} = 98.833$, and

$$m = \frac{3288.6 - 6(5.367)(98.833)}{187.36 - 6(5.367)^2} = 7.293$$

Thus

$$y' = 98.833 + 7.293(x - 5.367) = 7.293x + 59.691.$$

b. Predict the total sales if $5000 is spent on advertising.

**Answer:** Using the equation of the regression line, the resulting total sales will be $7.293(5) + 59.691 = 96.156$ thousands of dollars ($96,156).

c. Interpret the slope.

**Answer:** The slope of the regression line indicates that every extra $1000 spent on advertising will result in $7293 of added sales.

# EXERCISES

1. As reported in *The New York Times* (September 21, 1994, page C10), a study at the University of Toronto determined that, for every 10 grams of saturated fat consumed per day, a woman's risk of ovarian cancer rises 20%. What is the slope of the appropriate regression line, and what does it mean?

2. Jay Bennett (*Chance*, Winter 1995, page 38) calculated the regression line for average 1991 SAT scores (total math plus verbal) versus number of dollars spent per student in 1991 for New Jersey school districts and obtained a slope of 0.0227 and a y-intercept of 707.
   a. Find the equation of the regression line.
   b. Interpret this slope in terms of each additional $1000 spent per student.
   c. What average SAT result does this regression line predict for students in a district that spends $10,000 per student?
   d. According to this analysis, how much should a district spend per student in order for its students to average 1000 on the SAT exam?

3. According to *The New York Times* (April 2, 1993, page A1), the average monthly rate for basic television cable service has increased as follows:

| Year: | 1986 | 1987 | 1988 | 1989 | 1990 | 1991 | 1992 |
|---|---|---|---|---|---|---|---|
| Rate: | $11.00 | 13.20 | 13.90 | 15.20 | 16.80 | 18.00 | 20.00 |

a. Find the equation of the best-fitting straight line.
b. Interpret the slope.
c. Predict the average monthly rate in 1993 (actually it was $21.00).
d. Predict in what year the rate will reach $50.00.

4. The heart disease death rates per 100,000 people in the United States for certain years, as reported by the *National Center for Health Statistics*, were as follows:

| Year: | 1950 | 1960 | 1970 | 1975 | 1980 |
|---|---|---|---|---|---|
| Death rate: | 307.6 | 286.2 | 253.6 | 217.8 | 202.0 |

a. Find the equation of the best-fitting straight line.
b. Interpret the slope.
c. Predict the death rate in 1983 (actually it was 188.5).
d. Predict in what year the death rate will drop to 100.

5. In studying the "ozone-deficit" problem (*Science*, September 23, 1994, page 1835), one experiment measured laser-induced fluorescence (LIF) at various ozone partial pressures, obtaining the following results:

| Ozone partial pressure (mtorr): | 8 | 16 | 27 | 41 | 52 | 67 | 83 |
|---|---|---|---|---|---|---|---|
| LIF intensity: | 1.5 | 3 | 4.2 | 7.5 | 9.8 | 11.5 | 15 |

a. Find the equation of the best-fitting straight line.
b. Interpret the slope.
c. Predict the LIF intensity if the ozone partial pressure is 37 millitorr.
d. Predict what ozone partial pressure would result if the LIF intensity was 18.

## Study Question

Suppose we wish to minimize

$$\Sigma(y - y')^2 = (y_1 - y_1')^2 + (y_2 - y_2')^2 + \ldots + (y_n - y_n')^2$$
$$\text{where} \quad y_i' = \bar{y} + m(x_i - \bar{x})$$

We consider

$$\Sigma(y - [\bar{y} + m(x - \bar{x})])^2 = \Sigma[y - \bar{y} - m(x - \bar{x})]^2$$
$$= \Sigma(y - \bar{y})^2 - 2m \Sigma(y - \bar{y})(x - \bar{x})$$
$$+ m^2 \Sigma(x - \bar{x})^2$$

Using the fact that $am^2 + bm + c$ has its lowest point when $m = -b/2a$, show that we want

$$m = \frac{\Sigma(y - \bar{y})(x - \bar{x})}{\Sigma(x - \bar{x})^2}$$

Expand to

$$m = \frac{\Sigma\,xy - \Sigma\,x\bar{y} - \Sigma\,\bar{x}y + \Sigma\,\bar{x}\bar{y})}{\Sigma\,x^2 - 2\,\Sigma\,x\bar{x} + \Sigma\,\bar{x}^2}$$

Now, using the fact that $\bar{x}$ and $\bar{y}$ do not vary and so can be factored out of summations (e.g., $\Sigma x\bar{y} = \bar{y}\Sigma x$), and the fact that $\Sigma x = n\bar{x}$ and $\Sigma y = n\bar{y}$, show that the above expression for $m$ can be algebraically reduced to

$$m = \frac{\Sigma\,xy - n\bar{x}\bar{y}}{\Sigma\,x^2 - n(\bar{x})^2}$$

# 22
# CORRELATION

There are ways to gauge whether the perceived relationship between variables is strong enough so that finding the regression line and making use of it are meaningful. Although a scatter diagram will usually give an intuitive visual indication when a linear relationship is strong, in most cases it is quite difficult to visually judge the specific strength of a relationship. For this reason we develop a mathematical measure called the *correlation*. Important as correlation is, we need always to keep in mind that significant correlation does not necessarily indicate *causation*.

**Example 22.1**  Consider the data and results from Example 21.1. How good a fit is $y' = 2.548x - 52.16$? For example, compare the actual values to the values predicted by the regression formula. Plugging in the various values of $x$ gives $2.548(38) - 52.16 = 44.664$, $2.548(53) - 52.16 = 82.884$, $2.548(42) - 52.16 = 54.856$, $2.548(47) = 67.596$.

| x | y | y' |
|----|----|--------|
| 38 | 45 | 44.664 |
| 53 | 86 | 82.884 |
| 42 | 58 | 54.856 |
| 47 | 61 | 67.596 |

As tabulated above, the observed $y$-values and regression $y'$-values appear to be close. However, are they close enough to be significant?

# CORRELATION COEFFICIENT

One measure of an apparent relationship, such as the one above, is called the *correlation coefficient* and is denoted as $r$. Actually, $r$ is related to the $F$-ratio studied in Chapter 15. Like the $F$-ratio, $r$ is the square root of a ratio of variances.

$$r^2 = \frac{\sigma_{y'}^2}{\sigma_y^2}$$

where

$$\sigma^2_{y'} = \frac{\Sigma(y' - \bar{y})^2}{n} \qquad \text{and} \qquad \sigma^2_{y} = \frac{\Sigma(y - \bar{y})^2}{n}$$

In other words, $r^2$, called the *coefficient of determination*, is the ratio of the variance of the predicted values, $y'$, to the variance of the observed values, $y$. Alternatively, we can say that there is a partition of the $y$-variance, and $r^2$ is the proportion of this variance that is predictable from a knowledge of $x$.

Continuing Example 22.1, we calculate the variances as follows. From an earlier calculation we have $\bar{y} = 62.5$, and so

$$\sigma^2_{y} = \frac{(44.664 - 62.5)^2 + (82.884 - 62.5)^2 + (54.856 - 62.5)^2 + (67.596 - 62.5)}{4}$$

$$= \frac{818.03}{4} = 204.508 \text{ and}$$

$$\sigma^2_{y'} = \frac{(45 - 62.5)^2 + (86 - 62.5)^2 + (58 - 62.5)^2 + (61 - 62.5)^2}{4} = 220.25$$

Thus

$$r^2 = \frac{204.508}{220.25} = .9285$$

and $r = .9636$

It can be shown algebraically that

$$r^2 = \frac{(\Sigma xy - n\bar{x}\bar{y})^2}{\left[\Sigma x^2 - n(\bar{x})^2\right]\left[\Sigma y^2 - n(\bar{y})^2\right]}$$

The correlation coefficient $r$ will take the same sign as $m$ (the slope of the regression line). The value of $r$ always falls between $-1$ and $+1$, with $-1$ indicating perfect negative correlation, and $+1$ indicating perfect positive correlation.

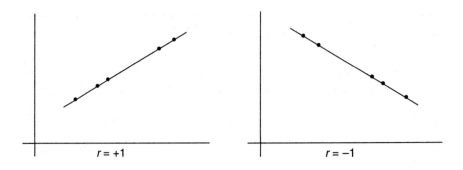

$r = +1$            $r = -1$

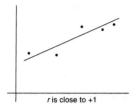

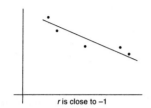

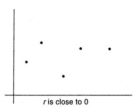

*r* is close to +1          *r* is close to –1          *r* is close to 0

# HYPOTHESIS TEST FOR CORRELATION

**Example 22.2**   Consider Example 21.2. How good a fit is $y' = 0.5492x - 1.73$? Plugging in the various values of $x$ to compare the actual values to the values predicted by the regression formula gives these results:

| x | y | y' |
|---|---|----|
| 25 | 10 | 12.00 |
| 23 | 11 | 10.90 |
| 30 | 14 | 14.75 |
| 25 | 12 | 12.00 |
| 20 | 8 | 9.25 |
| 33 | 18 | 16.39 |
| 18 | 9 | 8.16 |
| 21 | 10 | 9.80 |
| 22 | 10 | 10.35 |
| 30 | 15 | 14.75 |
| 26 | 11 | 12.55 |
| 26 | 15 | 12.55 |
| 27 | 12 | 13.10 |
| 29 | 14 | 14.20 |
| 20 | 11 | 9.25 |

Are the observed $y$ values and the regression $y'$ values close enough to be significant? We noted earlier that $n = 15$, $\Sigma x = 375$, $\bar{x} = 25$, $\Sigma y = 180$, $\bar{y} = 12$, $\Sigma x^2 = 9639$, and $\Sigma xy = 4645$. We now calculate

$$\Sigma y^2 = 10^2 + 11^2 + 14^2 + \ldots + 11^2 = 2262,$$

and thus

$$r^2 = \frac{(4645 - 15(25)(12))^2}{(9639 - 15(25)^2)(2262 - 15(12)^2)} = \frac{(145)^2}{(264)(102)} = .7808$$

and $r = .8836$.

Is $r = .9636$ with four data pairs significant? How about .8836 with 15 data pairs? If there really is no correlation, what are the chances that the $r$-values could have been this large? As expected, given our earlier hypothesis testing techniques, these answers depend on the level of significance with which we are working. Table E in the Appendix lists of critical $r$-values for both the 5% and 1% levels of significance; that is, with no correlation, there are .05 and .01 probabilities of obtaining the given critical levels. Thus, as before, we talk about $\alpha$-risks of .05 and .01, respectively. Note that to use Table E, the number of degrees of freedom must be known—in this situation $df = n - 2$.

We have

$$H_o: \text{no correlation}$$

$$H_a: \text{correlation}$$

If our calculated $r$, in absolute value, is greater than the critical $r$, we reject $H_0$ and say that at the given significance level there is sufficient evidence that $r$, the correlation coefficient, is significant.

For Example 22.1, $df = 4 - 2 = 2$, which gives critical values of $r$ of .950 and .990 at the 5% and 1% significance levels, respectively; $.9636 > .950$, but $.9636 < .990$. Thus, if we are willing to accept a 5% probability of committing a Type I error, we can conclude that there *is* sufficient evidence to indicate correlation. However, if we are willing to accept only a 1% probability of committing a Type I error, there is *not* sufficient evidence to reject the null hypothesis.

For Example 22.2, $df = 15 - 2 = 13$, which gives critical values of .514 and .641 at the 5% and 1% significance levels, respectively; $.8836 > .514$, and $.8836 > .641$. Therefore, at either of these significance levels, we conclude that there *is* sufficient evidence to indicate correlation.

**Example 22.3**  Following are the lengths and grades of ten research papers for a sociology professor's class.

| Length (pages): | $x$ | 25 | 32 | 20 | 28 | 15 | 34 | 29 | 30 | 45 | 35 |
|---|---|---|---|---|---|---|---|---|---|---|---|
| Grade: | $y$ | 69 | 81 | 72 | 75 | 64 | 89 | 84 | 73 | 92 | 86 |

a. Find the equation of the regression line.

***Answer:*** We calculate as follows:

$$\Sigma x = 25 + 32 + \ldots + 35 \qquad\qquad = 293$$
$$\Sigma y = 69 + 81 + \ldots + 86 \qquad\qquad = 785$$
$$\Sigma xy = 25(69) + 32(81) + \ldots + 35(86) = 23{,}619$$
$$\Sigma x^2 = 25^2 + 32^2 + \ldots + 35^2 \qquad\quad = 9{,}205$$
$$\Sigma y^2 = 69^2 + 81^2 + \ldots + 86^2 \qquad\quad = 62{,}393$$
$$\bar{x} = \frac{293}{10} = 29.3, \; \bar{y} = \frac{785}{10} = 78.5$$
$$m = \frac{23{,}619 - 10(29.3)(78.5)}{9205 - 10(29.3)^2} = \frac{618.5}{620.1} = 0.997$$
$$y' = 78.5 + 0.997(x - 29.3) = 0.997x + 49.3$$

b. Plot a scatter diagram and graph the regression line.

***Answer:***

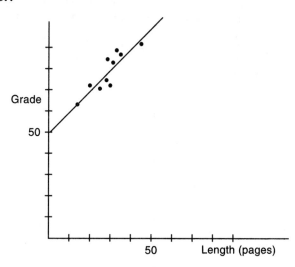

c. Use the equation to predict the grade for a student who turns in a paper 40 pages long.

***Answer:*** The grade for a student who turns in a paper 40 pages long is predicted to be $0.997(40) + 49.3 = 89.2$.

d. What is the slope of the regression line, and what does it signify?

***Answer:*** The slope of the regression line indicates that, for each additional page, students can increase their grades by 0.997, that is, approximately 1 point per page.

e. Test for correlation at both the 5% and 1% levels of significance.

***Answer:*** We calculate as follows:

$$r^2 = \frac{(618.5)^2}{(620.1)(62{,}393 - 10(78.5)^2)} = .801$$

and     $r = .895$

With $df = 10 - 2 = 8$, the critical values are .632 and .765 at the 5% and 1% significance levels, respectively; $.895 > .632$ and $.895 > .765$. Therefore, at either of these significance levels we conclude that there *is* sufficient evidence to indicate correlation between a student's grade and the number of pages in the research paper.

**Example 22.4**  The shoe sizes and the number of ties owned by ten corporate vice presidents are as follows.

| Shoe size: | $x$ | 8 | 9.5 | 9 | 11 | 9 | 9.5 | 8.5 | 9 | 9 | 9.5 |
| Number of ties: | $y$ | 10 | 10 | 8 | 15 | 12 | 13 | 16 | 7 | 12 | 4 |

a. Draw a scatter diagram for this data.

***Answer:***

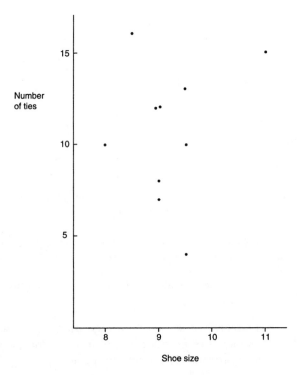

b. Test for correlation with $\alpha = .05$.

**Answer:** We calculate as follows:

$$\Sigma x = 8 + 9.5 + \ldots + 9.5 \qquad\qquad = 92$$

$$\Sigma y = 10 + 10 + \ldots + 4 \qquad\qquad = 107$$

$$\Sigma xy = 8(10) + 9.5(10) + \ldots + 9.5(4) = 988.5$$

$$\Sigma x^2 = 8^2 + 9.5^2 + \ldots + 9.5^2 \qquad = 852$$

$$\Sigma y^2 = 10^2 + 10^2 + \ldots + 4^2 \qquad = 1{,}267$$

$$\bar{x} = \frac{92}{10} = 9.2, \quad \bar{y} = \frac{107}{10} = 10.7$$

$$r^2 = \frac{(988.5 - 10(9.2)(10.7))^2}{(852 - 10(9.2)^2)(1267 - 10(10.7)^2)} = .02458$$

$$r = .1568$$

With $df = 10 - 2 = 8$ the critical $r$ value at the 5% level of significance is .632. Since $.1568 < .632$, we conclude there is not evidence of correlation.

c. Can we use the techniques of this chapter to find the best fitting straight line approximation to the above data? Does it make sense to use this equation to predict the number of ties owned by a corporate executive who wears size 10 shoes?

**Answer:** Although theoretically we can use our techniques to find the best fitting straight line approximation, the result will be meaningless and should not be used for predictions.

# EXERCISES

1. As reported in the *Journal of the American Medical Association* (June 13, 1990, page 3031), for a study of 10 nonagenarians (subjects were age 90 ± 1), the following tabulation shows a measure of strength (heaviest weight subject could lift using knee extensors) versus a measure of functional mobility (time taken to walk 6 meters).

| Strength (kg): | 7.5 | 6 | 11.5 | 10.5 | 9.5 | 18 | 4 | 12 | 9 | 3 |
|---|---|---|---|---|---|---|---|---|---|---|
| Walk time (s): | 18 | 46 | 8 | 25 | 25 | 7 | 22 | 12 | 10 | 48 |

a. Find the equation of the regression line.
b. Find the correlation coefficient, and test for correlation at the 5% significance level.

c. What is the sign of the slope of the regression line, and what does it signify?

2. The following table (*Statistical Abstract of the United States*) gives the U.S. population (in millions) in various years of the twentieth century.

| | | | |
|------|-------|------|-------|
| 1900 | 76.1  | 1950 | 151.9 |
| 1905 | 83.8  | 1955 | 165.1 |
| 1910 | 92.4  | 1960 | 180.0 |
| 1915 | 100.5 | 1965 | 193.5 |
| 1920 | 106.5 | 1970 | 204.0 |
| 1925 | 115.8 | 1975 | 215.5 |
| 1930 | 123.1 | 1980 | 227.2 |
| 1935 | 127.3 | 1985 | 237.9 |
| 1940 | 132.5 | 1990 | 249.4 |
| 1945 | 133.4 | | |

a. Find the equation of the regression line.
b. Find the correlation coefficient, and test for correlation at the 1% significance level.
c. Predict what the population will be in the year 2100.
d. What is the slope of the regression line, and what does it signify?
e. Predict when the population will reach 300 million.

3. In modeling population growth, an exponential function often gives a better correlation than a linear function.

a. Using the data from Exercise 2, make a table of years since 1900 versus the natural log of the above population values.
b. Find the equation of the regression line of ln $y$ in terms of $x$ (and calculate the correlation coefficient), and then solve for $y$ in terms of $x$. Your result will have the form

$$y = y_0 e^{kx}$$

where $y$ is the population (in millions) at a time $x$ years since 1900.
c. Predict what the population will be in the year 2100.
d. Predict when the population will reach 300 million.

4. The following table (*Statistical Abstract of the United States*) gives personal income in various years:

| | | | |
|------|--------|------|---------|
| 1935 | $ 467  | 1965 | $2,845  |
| 1940 | 586    | 1970 | 4,052   |
| 1945 | 1,215  | 1975 | 6,053   |
| 1950 | 1,502  | 1980 | 9,948   |
| 1955 | 1,903  | 1985 | 14,170  |
| 1960 | 2,264  | 1990 | 18,699  |

a. Find the equation of the regression line.
b. Find the correlation coefficient, and test for correlation at the 5% significance level.
c. Predict what the personal income will be in the year 2050.
d. What is the slope of the regression line, and what does it signify?
e. Predict when personal income will reach $25,000.

5. a. Using the data from Exercise 4, make a table of years since 1935 versus the natural log of the above personal income values.
b. Find the equation of the regression line of ln $y$ in terms of $x$, and calculate the correlation coefficient.
c. Solve for $y$ in terms of $x$.
d. Predict what the personal income will be in the year 2050.
e. Predict when personal income will reach $25,000.

## Study Questions

1. Find an example for which $r = 0$. Plot the scatter diagram.

2. a. Given that

$$r^2 = \frac{\sigma_{y'}^2}{\sigma_y^2} = \frac{\Sigma(y' - \bar{y})^2/n}{\Sigma(y - \bar{y})^2/n} = \frac{\Sigma(y' - \bar{y})^2}{\Sigma(y - \bar{y})^2}$$

and that $y' = \bar{y} + m(x - \bar{x})$, show that

$$r^2 = \frac{m^2 \Sigma(x - \bar{x})^2}{\Sigma(y - \bar{y})^2}$$

b. Now, using from the Study Question in Chapter 21 the equation

$$m = \frac{\Sigma(y - \bar{y})(x - \bar{x})}{\Sigma(x - \bar{x})^2}$$

show that     $r^2 = \dfrac{[\Sigma(y - \bar{y})(x - \bar{x})]^2}{\Sigma(x - \bar{x})^2 \ \Sigma(y - \bar{y})^2}$

c. Using expansion and algebraic techniques derive that

$$r^2 = \frac{(\Sigma xy - n\bar{x}\bar{y})^2}{\left[\Sigma x^2 - n(\bar{x})^2\right]\left[\Sigma y^2 - n(\bar{y})^2\right]}$$

3. An alternative approach is to standardize all scores, that is, to replace $x$ by

$$z_x = \frac{x - \bar{x}}{\sigma_x}$$

where

$$\sigma_x = \frac{\Sigma(x - \bar{x})^2}{n},$$

and to replace $y$ by

$$z_y = \frac{y - \bar{y}}{\sigma_y}$$

where

$$\sigma_y = \frac{\Sigma(y - \bar{y})^2}{n}.$$

If there were complete positive linear dependence of $y$ on $x$, then, for each standardized change in $x$, there would be an identical standardized change in $y$; that is, $z_y$ would equal $z_x$, for all $x$ and $y$, and $\frac{z_y}{z_x}$ would equal 1. More generally, if

$$z_{y'} = \frac{y' - \bar{y}}{\sigma_y}$$

then the correlation coefficient, $r$, can be defined as the slope $r = \frac{z_{y'}}{z_x}$.

a. Starting with $z_{y'} = rz_x$ use the properties of variances and the fact that $\sigma^2_{z_x} = 1$ to show that

$$\sigma^2_{z_{y'}} = r^2$$

b. Starting with $z_{y'} = \frac{y' - \bar{y}}{\sigma_y}$, or $y' = \sigma_y z_{y'} + \bar{y}$, use the properties of variances to show that

$$\sigma^2_{y'} = (\sigma_y)^2 \sigma^2_{z_{y'}}$$

c. Combine the results of parts $a$ and $b$ and conclude that $r^2 = \dfrac{\sigma^2_{y'}}{\sigma^2_y}$.

4. In this chapter we defined the square of the correlation coefficient to be a ratio of the variance $\dfrac{\Sigma (y' - \bar{y})^2}{n}$ to the variance $\dfrac{\Sigma (y - \bar{y})^2}{n}$. Here, $\Sigma(y' - \bar{y})^2$ is called the *explained variation*, while $\Sigma(y - \bar{y})^2$ is the *total variation*.

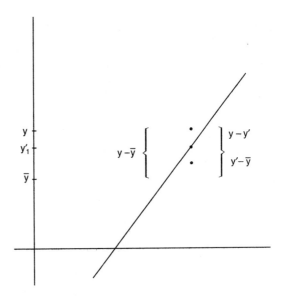

In Example 22.1 we found that the explained variation $\Sigma(y' - \bar{y})^2 = 818$, while the total variation $\Sigma(y - \bar{y})^2 = 881$.

a. Calculate $\Sigma(y - y')^2$, called the *unexplained variation*, and show that it is equal to the total variation, 881, minus the explained variation, 818.

b. Calculate the explained variation, the unexplained variation, and the total variation in Example 22.2.

# *APPENDICES*

In the real world, time and cost considerations usually make it impossible to analyze an entire population. Does the government question you and your parents before announcing the monthly unemployment rates? Does a producer check every household's viewing preferences before deciding whether a pilot program will continue? In studying statistics we learn how to estimate *population* characteristics by considering a *sample*. For example, in this book we estimate population means and proportions by looking at sample means and proportions.

To derive conclusions about the larger population, we need to be confident that the sample we have chosen fairly represents that population. Analyzing the data with computers is often easier then gathering the data, but the frequently quoted "garbage in, garbage out" applies here. Nothing can help if the data are badly collected. Unfortunately, many of the statistics with which we are bombarded through newspapers, radio, and television are based on poorly designed data-collection procedures.

Two common techniques that often result in flawed conclusions are *voluntary response samples* and *convenience samples*. Voluntary response samples, based on individuals who offer to participate, typically give too much emphasis to persons with strong opinions. For example, radio call-in programs about controversial topics such as gun control, abortion, and school segregation do not produce meaningful data on what proportion of the population favor or oppose related issues. Convenience samples, based on choosing individuals who are easy to reach, are also suspect. For example, interviews at shopping malls tend to produce data highly unrepresentative of the entire population.

Poorly designed sampling techniques result in *bias*, that is, in a tendency to favor the selection of certain members of a population. An often-cited example of *selection bias* is the *Literary Digest* opinion poll that predicted a landslide victory for Alfred Landon over Franklin D. Roosevelt in the 1936 presidential election. The *Digest* surveyed people with cars and telephones, but in 1936 only the wealthy minority, who mainly voted Republican, had cars and telephones. Examples of *nonresponse bias* are most mailed questionnaires. They tend to have very low response percentages, and it is often unclear which part of the population is responding. *Unintentional bias* often creeps in when the surveyor tries to systematically pick people representative of the whole population. For example, in 1948 the *Chicago Tribune* incorrectly called Dewey the winner over Truman. Here the mistake was partly due to misleading polls based on quota sampling which left the interviewers too much free choice in picking people to fill their quotas.

How can a good, that is, a *representative*, sample be chosen? The most accurate technique would be to write the name of each member of the population on a card, mix the cards thoroughly in a large box, and then pull out some specified number of these cards. This method would give everyone in the population an equal chance of being selected as part of the sample. Unfortunately, this method is usually too time consuming and too costly, and bias might still creep in if the mixing was not thorough. A *simple random sample* can more accurately be obtained by assigning a number to everyone in the population and then having a computer generate random numbers to indicate choices.

Time and cost-saving modifications are needed to implement sampling procedures. *Systematic sampling* involves randomly listing the population in order, and then picking every tenth, hundredth, or thousandth, etc., person from the list. In *stratified sampling* the population is divided into representative groups called *strata*, and random samples of persons from all strata are chosen. *Multistage sampling* involves dividing the population into groupings, then each grouping is subdivided, a random sampling of the subdivisions is selected, and finally a random sample of people is picked from the selected subdivisions. What is crucial in all these techniques is the presence of a systematic procedure that involves the use of *chance* and leaves no freedom of choice to the interviewers.

In statistical studies that aim to confirm relationships between variables, we must be cognizant of the difference between *observational* versus *controlled* studies. In observational studies there is no choice in regard to who goes into which group. For example, a researcher cannot ethically tell 100 people to smoke three packs of cigarettes a day and 100 others to smoke only one pack per day; he can only observe people who habitually smoke those amounts. In a controlled study, however, the researcher can randomly divide subjects into appropriate groups. For example, patients may be randomly given unmarked capsules of either aspirin or acetaminophen and then the effects measured. Controlled studies often have a *treatment group* and a *control group*; in the ideal situation, neither the subjects nor their attendants know which group is which. The Salk-vaccine experiment of the 1950s, in which half the children received the vaccine and half were given a placebo, with not even their doctors knowing who received what, is a classic example of this *double-blind* approach.

# IMPORTANT STATISTICAL TABLES

## TABLE A. NORMAL CURVE AREAS

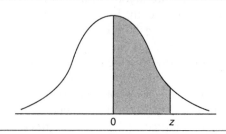

| z | .00 | .01 | .02 | .03 | .04 | .05 | .06 | .07 | .08 | .09 |
|-----|------|------|------|------|------|------|------|------|------|------|
| 0.0 | .0000 | .0040 | .0080 | .0120 | .0160 | .0199 | .0239 | .0279 | .0319 | .0359 |
| 0.1 | .0398 | .0438 | .0478 | .0517 | .0557 | .0596 | .0636 | .0675 | .0714 | .0753 |
| 0.2 | .0793 | .0832 | .0871 | .0910 | .0948 | .0987 | .1026 | .1064 | .1103 | .1141 |
| 0.3 | .1179 | .1217 | .1255 | .1293 | .1331 | .1368 | .1406 | .1443 | .1480 | .1517 |
| 0.4 | .1554 | .1591 | .1628 | .1664 | .1700 | .1736 | .1772 | .1808 | .1844 | .1879 |
| 0.5 | .1915 | .1950 | .1985 | .2019 | .2054 | .2088 | .2123 | .2157 | .2190 | .2224 |
| 0.6 | .2257 | .2291 | .2324 | .2357 | .2389 | .2422 | .2454 | .2486 | .2517 | .2549 |
| 0.7 | .2580 | .2611 | .2642 | .2673 | .2704 | .2734 | .2764 | .2794 | .2823 | .2852 |
| 0.8 | .2881 | .2910 | .2939 | .2967 | .2995 | .3023 | .3051 | .3078 | .3106 | .3133 |
| 0.9 | .3159 | .3186 | .3212 | .3238 | .3264 | .3289 | .3315 | .3340 | .3365 | .3389 |
| 1.0 | .3413 | .3438 | .3461 | .3485 | .3508 | .3531 | .3554 | .3577 | .3599 | .3621 |
| 1.1 | .3643 | .3665 | .3686 | .3708 | .3729 | .3749 | .3770 | .3790 | .3810 | .3830 |
| 1.2 | .3849 | .3869 | .3888 | .3907 | .3925 | .3944 | .3962 | .3980 | .3997 | .4015 |
| 1.3 | .4032 | .4049 | .4066 | .4082 | .4099 | .4115 | .4131 | .4147 | .4162 | .4177 |
| 1.4 | .4192 | .4207 | .4222 | .4236 | .4251 | .4265 | .4279 | .4292 | .4306 | .4319 |
| 1.5 | .4332 | .4345 | .4357 | .4370 | .4382 | .4394 | .4406 | .4418 | .4429 | .4441 |
| 1.6 | .4452 | .4463 | .4474 | .4484 | .4495 | .4505 | .4515 | .4525 | .4535 | .4545 |
| 1.7 | .4554 | .4564 | .4573 | .4582 | .4591 | .4599 | .4608 | .4616 | .4625 | .4633 |
| 1.8 | .4641 | .4649 | .4656 | .4664 | .4671 | .4678 | .4686 | .4693 | .4699 | .4706 |
| 1.9 | .4713 | .4719 | .4726 | .4732 | .4738 | .4744 | .4750 | .4756 | .4761 | .4767 |
| 2.0 | .4772 | .4778 | .4783 | .4788 | .4793 | .4798 | .4803 | .4808 | .4812 | .4817 |
| 2.1 | .4821 | .4826 | .4830 | .4834 | .4838 | .4842 | .4846 | .4850 | .4854 | .4857 |
| 2.2 | .4861 | .4864 | .4868 | .4871 | .4875 | .4878 | .4881 | .4884 | .4887 | .4890 |
| 2.3 | .4893 | .4896 | .4898 | .4901 | .4904 | .4906 | .4909 | .4911 | .4913 | .4916 |
| 2.4 | .4918 | .4920 | .4922 | .4925 | .4927 | .4929 | .4931 | .4932 | .4934 | .4936 |
| 2.5 | .4938 | .4940 | .4941 | .4943 | .4945 | .4946 | .4948 | .4949 | .4951 | .4952 |
| 2.6 | .4953 | .4955 | .4956 | .4957 | .4959 | .4960 | .4961 | .4962 | .4963 | .4964 |
| 2.7 | .4965 | .4966 | .4967 | .4968 | .4969 | .4970 | .4971 | .4972 | .4973 | .4974 |
| 2.8 | .4974 | .4975 | .4976 | .4977 | .4977 | .4978 | .4979 | .4979 | .4980 | .4981 |
| 2.9 | .4981 | .4982 | .4982 | .4983 | .4984 | .4984 | .4985 | .4985 | .4986 | .4986 |
| 3.0 | .4987 | .4987 | .4987 | .4988 | .4988 | .4989 | .4989 | .4989 | .4990 | .4990 |

Abridged from Table I of A. Hald, *Statistical Tables and Formulas* (New York: John Wiley & Sons, Inc.), 1952. Reproduced by permission of A. Hald and the publisher, John Wiley & Sons, Inc.

## TABLE B. CRITICAL VALUES OF *t*

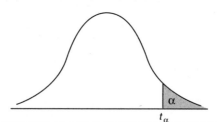

| df | $t_{.100}$ | $t_{.050}$ | $t_{.025}$ | $t_{.010}$ | $t_{.005}$ | $t_{.001}$ | $t_{.0005}$ |
|----|------|------|------|------|------|------|------|
| 1 | 3.078 | 6.314 | 12.706 | 31.821 | 63.657 | 318.31 | 636.62 |
| 2 | 1.886 | 2.920 | 4.303 | 6.965 | 9.925 | 22.326 | 31.598 |
| 3 | 1.638 | 2.353 | 3.182 | 4.541 | 5.841 | 10.213 | 12.924 |
| 4 | 1.533 | 2.132 | 2.776 | 3.747 | 4.604 | 7.173 | 8.610 |
| 5 | 1.476 | 2.015 | 2.571 | 3.365 | 4.032 | 5.893 | 6.869 |
| 6 | 1.440 | 1.943 | 2.447 | 3.143 | 3.707 | 5.208 | 5.959 |
| 7 | 1.415 | 1.895 | 2.365 | 2.998 | 3.499 | 4.785 | 5.408 |
| 8 | 1.397 | 1.860 | 2.306 | 2.896 | 3.355 | 4.501 | 5.041 |
| 9 | 1.383 | 1.833 | 2.262 | 2.821 | 3.250 | 4.297 | 4.781 |
| 10 | 1.372 | 1.812 | 2.228 | 2.764 | 3.169 | 4.144 | 4.587 |
| 11 | 1.363 | 1.796 | 2.201 | 2.718 | 3.106 | 4.025 | 4.437 |
| 12 | 1.356 | 1.782 | 2.179 | 2.681 | 3.055 | 3.930 | 4.318 |
| 13 | 1.350 | 1.771 | 2.160 | 2.650 | 3.012 | 3.852 | 4.221 |
| 14 | 1.345 | 1.761 | 2.145 | 2.624 | 2.977 | 3.787 | 4.140 |
| 15 | 1.341 | 1.753 | 2.131 | 2.602 | 2.947 | 3.733 | 4.073 |
| 16 | 1.337 | 1.746 | 2.120 | 2.583 | 2.921 | 3.686 | 4.015 |
| 17 | 1.333 | 1.740 | 2.110 | 2.567 | 2.898 | 3.646 | 3.965 |
| 18 | 1.330 | 1.734 | 2.101 | 2.552 | 2.878 | 3.610 | 3.922 |
| 19 | 1.328 | 1.729 | 2.093 | 2.539 | 2.861 | 3.579 | 3.883 |
| 20 | 1.325 | 1.725 | 2.086 | 2.528 | 2.845 | 3.552 | 3.850 |
| 21 | 1.323 | 1.721 | 2.080 | 2.518 | 2.831 | 3.527 | 3.819 |
| 22 | 1.321 | 1.717 | 2.074 | 2.508 | 2.819 | 3.505 | 3.792 |
| 23 | 1.319 | 1.714 | 2.069 | 2.500 | 2.807 | 3.485 | 3.767 |
| 24 | 1.318 | 1.711 | 2.064 | 2.492 | 2.797 | 3.467 | 3.745 |
| 25 | 1.316 | 1.708 | 2.060 | 2.485 | 2.787 | 3.450 | 3.725 |
| 26 | 1.315 | 1.706 | 2.056 | 2.479 | 2.779 | 3.435 | 3.707 |
| 27 | 1.314 | 1.703 | 2.052 | 2.473 | 2.771 | 3.421 | 3.690 |
| 28 | 1.313 | 1.701 | 2.048 | 2.467 | 2.763 | 3.408 | 3.674 |
| 29 | 1.311 | 1.699 | 2.045 | 2.462 | 2.756 | 3.396 | 3.659 |
| 30 | 1.310 | 1.697 | 2.042 | 2.457 | 2.750 | 3.385 | 3.646 |
| 40 | 1.303 | 1.684 | 2.021 | 2.423 | 2.704 | 3.307 | 3.551 |
| 60 | 1.296 | 1.671 | 2.000 | 2.390 | 2.660 | 3.232 | 3.460 |
| 120 | 1.289 | 1.658 | 1.980 | 2.358 | 2.617 | 3.160 | 3.373 |
| ∞ | 1.282 | 1.645 | 1.960 | 2.326 | 2.576 | 3.090 | 3.291 |

# TABLE C. CRITICAL VALUES OF F
## Values of F .05

**Denominator degrees of freedom**

**Numerator degrees of freedom**

| | 1 | 2 | 3 | 4 | 5 | 6 | 7 | 8 | 9 | 10 | 12 | 15 | 20 | 24 | 30 | 40 | 60 | 120 | ∞ |
|---|---|---|---|---|---|---|---|---|---|---|---|---|---|---|---|---|---|---|---|
| 1 | 161 | 200 | 216 | 225 | 230 | 234 | 237 | 239 | 241 | 242 | 244 | 246 | 248 | 249 | 250 | 251 | 252 | 253 | 254 |
| 2 | 18.5 | 19.0 | 19.2 | 19.2 | 19.3 | 19.3 | 19.4 | 19.4 | 19.4 | 19.4 | 19.4 | 19.4 | 19.4 | 19.5 | 19.5 | 19.5 | 19.5 | 19.5 | 19.5 |
| 3 | 10.1 | 9.55 | 9.28 | 9.12 | 9.01 | 8.94 | 8.89 | 8.85 | 8.81 | 8.79 | 8.74 | 8.70 | 8.66 | 8.64 | 8.62 | 8.59 | 8.57 | 8.55 | 8.53 |
| 4 | 7.71 | 6.94 | 6.59 | 6.39 | 6.26 | 6.16 | 6.09 | 6.04 | 6.00 | 5.96 | 5.91 | 5.86 | 5.80 | 5.77 | 5.75 | 5.72 | 5.69 | 5.66 | 5.63 |
| 5 | 6.61 | 5.79 | 5.41 | 5.19 | 5.05 | 4.95 | 4.88 | 4.82 | 4.77 | 4.74 | 4.68 | 4.62 | 4.56 | 4.53 | 4.50 | 4.46 | 4.43 | 4.40 | 4.37 |
| 6 | 5.99 | 5.14 | 4.76 | 4.53 | 4.39 | 4.28 | 4.21 | 4.15 | 4.10 | 4.06 | 4.00 | 3.94 | 3.87 | 3.84 | 3.81 | 3.77 | 3.74 | 3.70 | 3.67 |
| 7 | 5.59 | 4.74 | 4.35 | 4.12 | 3.97 | 3.87 | 3.79 | 3.73 | 3.68 | 3.64 | 3.57 | 3.51 | 3.44 | 3.41 | 3.38 | 3.34 | 3.30 | 3.27 | 3.23 |
| 8 | 5.32 | 4.46 | 4.07 | 3.84 | 3.69 | 3.58 | 3.50 | 3.44 | 3.39 | 3.35 | 3.28 | 3.22 | 3.15 | 3.12 | 3.08 | 3.04 | 3.01 | 2.97 | 2.93 |
| 9 | 5.12 | 4.26 | 3.86 | 3.63 | 3.48 | 3.37 | 3.29 | 3.23 | 3.18 | 3.14 | 3.07 | 3.01 | 2.94 | 2.90 | 2.86 | 2.83 | 2.79 | 2.75 | 2.71 |
| 10 | 4.96 | 4.10 | 3.71 | 3.48 | 3.33 | 3.22 | 3.14 | 3.07 | 3.02 | 2.98 | 2.91 | 2.85 | 2.77 | 2.74 | 2.70 | 2.66 | 2.62 | 2.58 | 2.54 |
| 11 | 4.84 | 3.98 | 3.59 | 3.36 | 3.20 | 3.09 | 3.01 | 2.95 | 2.90 | 2.85 | 2.79 | 2.72 | 2.65 | 2.61 | 2.57 | 2.53 | 2.49 | 2.45 | 2.40 |
| 12 | 4.75 | 3.89 | 3.49 | 3.26 | 3.11 | 3.00 | 2.91 | 2.85 | 2.80 | 2.75 | 2.69 | 2.62 | 2.54 | 2.51 | 2.47 | 2.43 | 2.38 | 2.34 | 2.30 |
| 13 | 4.67 | 3.81 | 3.41 | 3.18 | 3.03 | 2.92 | 2.83 | 2.77 | 2.71 | 2.67 | 2.60 | 2.53 | 2.46 | 2.42 | 2.38 | 2.34 | 2.30 | 2.25 | 2.21 |
| 14 | 4.60 | 3.74 | 3.34 | 3.11 | 2.96 | 2.85 | 2.76 | 2.70 | 2.65 | 2.60 | 2.53 | 2.46 | 2.39 | 2.35 | 2.31 | 2.27 | 2.22 | 2.18 | 2.13 |
| 15 | 4.54 | 3.68 | 3.29 | 3.06 | 2.90 | 2.79 | 2.71 | 2.64 | 2.59 | 2.54 | 2.48 | 2.40 | 2.33 | 2.29 | 2.25 | 2.20 | 2.16 | 2.11 | 2.07 |

0.05

$F_{.05}$

0

| | | | | | | | | | | | | | | | | | | | |
|---|---|---|---|---|---|---|---|---|---|---|---|---|---|---|---|---|---|---|---|
| 16 | 4.49 | 3.63 | 3.24 | 3.01 | 2.85 | 2.74 | 2.66 | 2.59 | 2.54 | 2.49 | 2.42 | 2.35 | 2.28 | 2.24 | 2.19 | 2.15 | 2.11 | 2.06 | 2.01 |
| 17 | 4.45 | 3.59 | 3.20 | 2.96 | 2.81 | 2.70 | 2.61 | 2.55 | 2.49 | 2.45 | 2.38 | 2.31 | 2.23 | 2.19 | 2.15 | 2.10 | 2.06 | 2.01 | 1.96 |
| 18 | 4.41 | 3.55 | 3.16 | 2.93 | 2.77 | 2.66 | 2.58 | 2.51 | 2.46 | 2.41 | 2.34 | 2.27 | 2.19 | 2.15 | 2.11 | 2.06 | 2.02 | 1.97 | 1.92 |
| 19 | 4.38 | 3.52 | 3.13 | 2.90 | 2.74 | 2.63 | 2.54 | 2.48 | 2.42 | 2.38 | 2.31 | 2.23 | 2.16 | 2.11 | 2.07 | 2.03 | 1.98 | 1.93 | 1.88 |
| 20 | 4.35 | 3.49 | 3.10 | 2.87 | 2.71 | 2.60 | 2.51 | 2.45 | 2.39 | 2.35 | 2.28 | 2.20 | 2.12 | 2.08 | 2.04 | 1.99 | 1.95 | 1.90 | 1.84 |
| 21 | 4.32 | 3.47 | 3.07 | 2.84 | 2.68 | 2.57 | 2.49 | 2.42 | 2.37 | 2.32 | 2.25 | 2.18 | 2.10 | 2.05 | 2.01 | 1.96 | 1.92 | 1.87 | 1.81 |
| 22 | 4.30 | 3.44 | 3.05 | 2.82 | 2.66 | 2.55 | 2.46 | 2.40 | 2.34 | 2.30 | 2.23 | 2.15 | 2.07 | 2.03 | 1.98 | 1.94 | 1.89 | 1.84 | 1.78 |
| 23 | 4.28 | 3.42 | 3.03 | 2.80 | 2.64 | 2.53 | 2.44 | 2.37 | 2.32 | 2.27 | 2.20 | 2.13 | 2.05 | 2.01 | 1.96 | 1.91 | 1.86 | 1.81 | 1.76 |
| 24 | 4.26 | 3.40 | 3.01 | 2.78 | 2.62 | 2.51 | 2.42 | 2.36 | 2.30 | 2.25 | 2.18 | 2.11 | 2.03 | 1.98 | 1.94 | 1.89 | 1.84 | 1.79 | 1.73 |
| 25 | 4.24 | 3.39 | 2.99 | 2.76 | 2.60 | 2.49 | 2.40 | 2.34 | 2.28 | 2.24 | 2.16 | 2.09 | 2.01 | 1.96 | 1.92 | 1.87 | 1.82 | 1.77 | 1.71 |
| 30 | 4.17 | 3.32 | 2.92 | 2.69 | 2.53 | 2.42 | 2.33 | 2.27 | 2.21 | 2.16 | 2.09 | 2.01 | 1.93 | 1.89 | 1.84 | 1.79 | 1.74 | 1.68 | 1.62 |
| 40 | 4.08 | 3.23 | 2.84 | 2.61 | 2.45 | 2.34 | 2.25 | 2.18 | 2.12 | 2.08 | 2.00 | 1.92 | 1.84 | 1.79 | 1.74 | 1.69 | 1.64 | 1.58 | 1.51 |
| 60 | 4.00 | 3.15 | 2.76 | 2.53 | 2.37 | 2.25 | 2.17 | 2.10 | 2.04 | 1.99 | 1.92 | 1.84 | 1.75 | 1.70 | 1.65 | 1.59 | 1.53 | 1.47 | 1.39 |
| 120 | 3.92 | 3.07 | 2.68 | 2.45 | 2.29 | 2.18 | 2.09 | 2.02 | 1.96 | 1.91 | 1.83 | 1.75 | 1.66 | 1.61 | 1.55 | 1.50 | 1.43 | 1.35 | 1.25 |
| ∞ | 3.84 | 3.00 | 2.60 | 2.37 | 2.21 | 2.10 | 2.01 | 1.94 | 1.88 | 1.83 | 1.75 | 1.67 | 1.57 | 1.52 | 1.46 | 1.39 | 1.32 | 1.22 | 1.00 |

**Denominator degrees of freedom**

## Values of $F_{.01}$

### Numerator degrees of freedom

| | 1 | 2 | 3 | 4 | 5 | 6 | 7 | 8 | 9 | 10 | 12 | 15 | 20 | 24 | 30 | 40 | 60 | 120 | ∞ |
|---|---|---|---|---|---|---|---|---|---|---|---|---|---|---|---|---|---|---|---|
| 1 | 4052 | 5000 | 5403 | 5625 | 5764 | 5859 | 5928 | 5982 | 6023 | 6056 | 6106 | 6157 | 6209 | 6235 | 6261 | 6287 | 6313 | 6339 | 6366 |
| 2 | 98.5 | 99.0 | 99.2 | 99.2 | 99.3 | 99.3 | 99.4 | 99.4 | 99.4 | 99.4 | 99.4 | 99.4 | 99.4 | 99.5 | 99.5 | 99.5 | 99.5 | 99.5 | 99.5 |
| 3 | 34.1 | 30.8 | 29.5 | 28.7 | 28.2 | 27.9 | 27.7 | 27.5 | 27.3 | 27.2 | 27.1 | 26.9 | 26.7 | 26.6 | 26.5 | 26.4 | 26.3 | 26.2 | 26.1 |
| 4 | 21.2 | 18.0 | 16.7 | 16.0 | 15.5 | 15.2 | 15.0 | 14.8 | 14.7 | 14.5 | 14.4 | 14.2 | 14.0 | 13.9 | 13.8 | 13.7 | 13.7 | 13.6 | 13.5 |
| 5 | 16.3 | 13.3 | 12.1 | 11.4 | 11.0 | 10.7 | 10.5 | 10.3 | 10.2 | 10.1 | 9.89 | 9.72 | 9.55 | 9.47 | 9.38 | 9.29 | 9.20 | 9.11 | 9.02 |
| 6 | 13.7 | 10.9 | 9.78 | 9.15 | 8.75 | 8.47 | 8.26 | 8.10 | 7.98 | 7.87 | 7.72 | 7.56 | 7.40 | 7.31 | 7.23 | 7.14 | 7.06 | 6.97 | 6.88 |
| 7 | 12.2 | 9.55 | 8.45 | 7.85 | 7.46 | 7.19 | 6.99 | 6.84 | 6.72 | 6.62 | 6.47 | 6.31 | 6.16 | 6.07 | 5.99 | 5.91 | 5.82 | 5.74 | 5.65 |
| 8 | 11.3 | 8.65 | 7.59 | 7.01 | 6.63 | 6.37 | 6.18 | 6.03 | 5.91 | 5.81 | 5.67 | 5.52 | 5.36 | 5.28 | 5.20 | 5.12 | 5.03 | 4.95 | 4.86 |
| 9 | 10.6 | 8.02 | 6.99 | 6.42 | 6.06 | 5.80 | 5.61 | 5.47 | 5.35 | 5.26 | 5.11 | 4.96 | 4.81 | 4.73 | 4.65 | 4.57 | 4.48 | 4.40 | 4.31 |
| 10 | 10.0 | 7.56 | 6.55 | 5.99 | 5.64 | 5.39 | 5.20 | 5.06 | 4.94 | 4.85 | 4.71 | 4.56 | 4.41 | 4.33 | 4.25 | 4.17 | 4.08 | 4.00 | 3.91 |
| 11 | 9.65 | 7.21 | 6.22 | 5.67 | 5.32 | 5.07 | 4.89 | 4.74 | 4.63 | 4.54 | 4.40 | 4.25 | 4.10 | 4.02 | 3.94 | 3.86 | 3.78 | 3.69 | 3.60 |
| 12 | 9.33 | 6.93 | 5.95 | 5.41 | 5.06 | 4.82 | 4.64 | 4.50 | 4.39 | 4.30 | 4.16 | 4.01 | 3.86 | 3.78 | 3.70 | 3.62 | 3.54 | 3.45 | 3.36 |
| 13 | 9.07 | 6.70 | 5.74 | 5.21 | 4.86 | 4.62 | 4.44 | 4.30 | 4.19 | 4.10 | 3.96 | 3.82 | 3.66 | 3.59 | 3.51 | 3.43 | 3.34 | 3.25 | 3.17 |
| 14 | 8.86 | 6.51 | 5.56 | 5.04 | 4.70 | 4.46 | 4.28 | 4.14 | 4.03 | 3.94 | 3.80 | 3.66 | 3.51 | 3.43 | 3.35 | 3.27 | 3.18 | 3.09 | 3.00 |
| 15 | 8.68 | 6.36 | 5.42 | 4.89 | 4.56 | 4.32 | 4.14 | 4.00 | 3.89 | 3.80 | 3.67 | 3.52 | 3.37 | 3.29 | 3.21 | 3.13 | 3.05 | 2.96 | 2.87 |

Denominator degrees of freedom

| df | | | | | | | | | | | | | | | | | | | |
|---|---|---|---|---|---|---|---|---|---|---|---|---|---|---|---|---|---|---|---|
| 16 | 8.53 | 6.23 | 5.29 | 4.77 | 4.44 | 4.20 | 4.03 | 3.89 | 3.78 | 3.69 | 3.55 | 3.41 | 3.26 | 3.18 | 3.10 | 3.02 | 2.93 | 2.84 | 2.75 |
| 17 | 8.40 | 6.11 | 5.19 | 4.67 | 4.34 | 4.10 | 3.93 | 3.79 | 3.68 | 3.59 | 3.46 | 3.31 | 3.16 | 3.08 | 3.00 | 2.92 | 2.83 | 2.75 | 2.65 |
| 18 | 8.29 | 6.01 | 5.09 | 4.58 | 4.25 | 4.01 | 3.84 | 3.71 | 3.60 | 3.51 | 3.37 | 3.23 | 3.08 | 3.00 | 2.92 | 2.84 | 2.75 | 2.66 | 2.57 |
| 19 | 8.19 | 5.93 | 5.01 | 4.50 | 4.17 | 3.94 | 3.77 | 3.63 | 3.52 | 3.43 | 3.30 | 3.15 | 3.00 | 2.92 | 2.84 | 2.76 | 2.67 | 2.58 | 2.49 |
| 20 | 8.10 | 5.85 | 4.94 | 4.43 | 4.10 | 3.87 | 3.70 | 3.56 | 3.46 | 3.37 | 3.23 | 3.09 | 2.94 | 2.86 | 2.78 | 2.69 | 2.61 | 2.52 | 2.42 |
| 21 | 8.02 | 5.78 | 4.87 | 4.37 | 4.04 | 3.81 | 3.64 | 3.51 | 3.40 | 3.31 | 3.17 | 3.03 | 2.88 | 2.80 | 2.72 | 2.64 | 2.55 | 2.46 | 2.36 |
| 22 | 7.95 | 5.72 | 4.82 | 4.31 | 3.99 | 3.76 | 3.59 | 3.45 | 3.35 | 3.26 | 3.12 | 2.98 | 2.83 | 2.75 | 2.67 | 2.58 | 2.50 | 2.40 | 2.31 |
| 23 | 7.88 | 5.66 | 4.76 | 4.26 | 3.94 | 3.71 | 3.54 | 3.41 | 3.30 | 3.21 | 3.07 | 2.93 | 2.78 | 2.70 | 2.62 | 2.54 | 2.45 | 2.35 | 2.26 |
| 24 | 7.82 | 5.61 | 4.72 | 4.22 | 3.90 | 3.67 | 3.50 | 3.36 | 3.26 | 3.17 | 3.03 | 2.89 | 2.74 | 2.66 | 2.58 | 2.48 | 2.40 | 2.31 | 2.21 |
| 25 | 7.77 | 5.57 | 4.68 | 4.18 | 3.86 | 3.63 | 3.36 | 3.32 | 3.22 | 3.13 | 2.99 | 2.85 | 2.70 | 2.62 | 2.53 | 2.45 | 2.36 | 2.27 | 2.17 |
| 30 | 7.56 | 5.39 | 4.51 | 4.02 | 3.70 | 3.47 | 3.30 | 3.17 | 3.07 | 2.98 | 2.84 | 2.70 | 2.55 | 2.47 | 2.39 | 2.30 | 2.21 | 2.11 | 2.01 |
| 40 | 7.31 | 5.18 | 4.31 | 3.83 | 3.51 | 3.29 | 3.12 | 2.99 | 2.89 | 2.80 | 2.66 | 2.52 | 2.37 | 2.29 | 2.20 | 2.11 | 2.02 | 1.92 | 1.80 |
| 60 | 7.08 | 4.98 | 4.13 | 3.65 | 3.34 | 3.12 | 2.95 | 2.82 | 2.72 | 2.63 | 2.50 | 2.35 | 2.20 | 2.12 | 2.03 | 1.94 | 1.84 | 1.73 | 1.60 |
| 120 | 6.85 | 4.79 | 3.95 | 3.48 | 3.17 | 2.96 | 2.79 | 2.66 | 2.56 | 2.47 | 2.34 | 2.19 | 2.03 | 1.95 | 1.86 | 1.76 | 1.66 | 1.53 | 1.38 |
| ∞ | 6.63 | 4.61 | 3.78 | 3.32 | 3.02 | 2.80 | 2.64 | 2.51 | 2.41 | 2.32 | 2.18 | 2.04 | 1.88 | 1.79 | 1.70 | 1.59 | 1.47 | 1.32 | 1.00 |

Denominator degrees of freedom

Reproduced from Table 18 of *Biometrika Tables for Statistics*, Vol. 1, 1966, by permission of the Biometrika Trustees.

## TABLE D. THE $\chi^2$-DISTRIBUTION

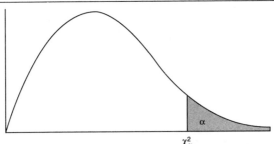

$\chi^2_\alpha$

| df | $\chi^2_{.995}$ | $\chi^2_{.990}$ | $\chi^2_{.975}$ | $\chi^2_{.950}$ | $\chi^2_{.900}$ | $\chi^2_{.100}$ | $\chi^2_{.050}$ | $\chi^2_{.025}$ | $\chi^2_{.010}$ | $\chi^2_{.005}$ |
|---|---|---|---|---|---|---|---|---|---|---|
| 1 | .0000 | .0002 | .0010 | .0039 | .0158 | 2.706 | 3.841 | 5.024 | 6.635 | 7.879 |
| 2 | .0100 | .0201 | .0506 | .1026 | .2107 | 4.605 | 5.991 | 7.378 | 9.210 | 10.60 |
| 3 | .0717 | .1148 | .2158 | .3518 | .5844 | 6.251 | 7.815 | 9.348 | 11.34 | 12.84 |
| 4 | .2070 | .2971 | .4844 | .7107 | 1.064 | 7.779 | 9.448 | 11.14 | 13.28 | 14.86 |
| 5 | .4117 | .5543 | .8312 | 1.145 | 1.610 | 9.236 | 11.07 | 12.83 | 15.09 | 16.75 |
| 6 | .6757 | .8721 | 1.237 | 1.635 | 2.204 | 10.64 | 12.59 | 14.45 | 16.81 | 18.55 |
| 7 | .9893 | 1.239 | 1.690 | 2.167 | 2.833 | 12.02 | 14.07 | 16.01 | 18.48 | 20.28 |
| 8 | 1.344 | 1.647 | 2.180 | 2.732 | 3.490 | 13.36 | 15.51 | 17.53 | 20.09 | 21.95 |
| 9 | 1.735 | 2.088 | 2.700 | 3.325 | 4.168 | 14.68 | 16.92 | 19.02 | 21.67 | 23.59 |
| 10 | 2.156 | 2.558 | 3.247 | 3.940 | 4.865 | 15.99 | 18.31 | 20.48 | 23.21 | 25.19 |
| 11 | 2.603 | 3.053 | 3.816 | 4.575 | 5.578 | 17.27 | 19.68 | 21.92 | 24.72 | 26.76 |
| 12 | 3.074 | 3.571 | 4.404 | 5.226 | 6.304 | 18.55 | 21.03 | 23.34 | 26.22 | 28.30 |
| 13 | 3.565 | 4.107 | 5.009 | 5.892 | 7.042 | 19.81 | 22.36 | 24.74 | 27.69 | 29.82 |
| 14 | 4.075 | 4.660 | 5.629 | 6.571 | 7.790 | 21.06 | 23.68 | 26.12 | 29.14 | 31.32 |
| 15 | 4.601 | 5.229 | 6.262 | 7.261 | 8.547 | 22.31 | 25.00 | 27.49 | 30.58 | 32.80 |
| 16 | 5.142 | 5.812 | 6.908 | 7.962 | 9.312 | 23.54 | 26.30 | 28.85 | 32.00 | 34.27 |
| 17 | 5.697 | 6.408 | 7.564 | 8.672 | 10.09 | 24.77 | 27.59 | 30.19 | 33.41 | 35.72 |
| 18 | 6.265 | 7.015 | 8.231 | 9.390 | 10.86 | 25.99 | 28.87 | 31.53 | 34.81 | 37.16 |
| 19 | 6.844 | 7.633 | 8.907 | 10.12 | 11.65 | 27.20 | 30.14 | 32.85 | 36.19 | 38.58 |
| 20 | 7.434 | 8.260 | 9.591 | 10.85 | 12.44 | 28.41 | 31.41 | 34.17 | 37.57 | 40.00 |
| 21 | 8.034 | 8.897 | 10.28 | 11.59 | 13.24 | 29.62 | 32.67 | 35.48 | 38.93 | 41.40 |
| 22 | 8.643 | 9.542 | 10.98 | 12.34 | 14.04 | 30.81 | 33.92 | 36.78 | 40.29 | 42.80 |
| 23 | 9.260 | 10.20 | 11.69 | 13.09 | 14.85 | 32.01 | 35.17 | 38.08 | 41.64 | 44.18 |
| 24 | 9.886 | 10.86 | 12.40 | 13.85 | 15.66 | 33.20 | 36.42 | 39.36 | 42.98 | 45.56 |
| 25 | 10.52 | 11.52 | 13.12 | 14.61 | 16.47 | 34.38 | 37.65 | 40.65 | 44.31 | 46.93 |
| 30 | 13.79 | 14.95 | 16.79 | 18.49 | 20.60 | 40.26 | 43.77 | 46.98 | 50.89 | 53.67 |
| 40 | 20.71 | 22.16 | 24.43 | 26.51 | 29.05 | 51.81 | 55.76 | 59.34 | 63.69 | 66.77 |
| 50 | 27.99 | 29.71 | 32.36 | 34.76 | 37.69 | 63.17 | 67.51 | 71.42 | 76.15 | 79.49 |
| 60 | 35.53 | 37.48 | 40.48 | 43.19 | 46.46 | 74.40 | 79.08 | 83.30 | 88.38 | 91.95 |
| 70 | 43.27 | 45.44 | 48.76 | 51.74 | 55.33 | 85.53 | 90.53 | 95.02 | 100.4 | 104.2 |
| 80 | 51.17 | 53.54 | 57.15 | 60.39 | 64.28 | 96.58 | 101.9 | 106.6 | 112.3 | 116.3 |
| 90 | 59.20 | 61.75 | 65.65 | 69.13 | 73.29 | 107.6 | 113.1 | 118.1 | 124.1 | 128.3 |
| 100 | 67.33 | 70.66 | 74.22 | 77.93 | 82.86 | 118.5 | 124.3 | 129.6 | 135.8 | 140.2 |

Adapted with permission from *Biometrika Tables for Statisticians*, Vol. 1. 3d ed., Cambridge University Press, 1966, edited by E. S. Pearson and H. O. Hartley.

## TABLE E. CRITICAL LEVELS OF *r* AT 5% AND 1% LEVELS OF SIGNIFICANCE

| df | $r_{.05}$ | $r_{.01}$ | df | $r_{.05}$ | $r_{.01}$ |
|---|---|---|---|---|---|
| 1 | .997 | 1.000 | 24 | .388 | .496 |
| 2 | .950 | .990 | 25 | .381 | .487 |
| 3 | .878 | .959 | 26 | .374 | .478 |
| 4 | .811 | .917 | 27 | .367 | .470 |
| 5 | .754 | .874 | 28 | .361 | .463 |
| 6 | .707 | .834 | 29 | .355 | .456 |
| 7 | .666 | .798 | 30 | .349 | .449 |
| 8 | .632 | .765 | 35 | .325 | .418 |
| 9 | .602 | .735 | 40 | .304 | .393 |
| 10 | .576 | .708 | 45 | .288 | .372 |
| 11 | .553 | .684 | 50 | .273 | .354 |
| 12 | .532 | .661 | 60 | .250 | .325 |
| 13 | .514 | .641 | 70 | .232 | .302 |
| 14 | .497 | .623 | 80 | .217 | .283 |
| 15 | .482 | .606 | 90 | .205 | .267 |
| 16 | .468 | .590 | 100 | .195 | .254 |
| 17 | .456 | .575 | 125 | .174 | .228 |
| 18 | .444 | .561 | 150 | .159 | .208 |
| 19 | .433 | .549 | 200 | .138 | .181 |
| 20 | .423 | .537 | 300 | .113 | .148 |
| 21 | .413 | .526 | 400 | .098 | .128 |
| 22 | .404 | .515 | 500 | .088 | .115 |
| 23 | .396 | .505 | 1000 | .062 | .081 |

Reproduced by the courtesy of the author and of the publisher from G. W. Snedecor and W. G. Cochran, *Statistical Methods.* The Iowa State University Press, Ames, Iowa, 1967, Table A11, p. 557.

# ANSWERS TO EXERCISES

## Chapter 1

1. The set is {0,0,0,0,0,0,0,0,0,0,0,1,1,1,1,1,1,1,1,3} so the mean is 11/19, the median is 1, and the mode is 0.

2. The median is $(11,187 + 9,333)/2 = 10,260$, the mean is 42,785.75, and the trimmed mean (with the upper and lower 25% removed) is 17,252.

3. From $(0.70 − 0.11 + 1.34 + 0.05 + x)/5 = 0.086$, the return was −1.55%.

4. a. 255 lb
   b. 265 lb
   c. 280.5 lb

5. The geometric mean of 31.07 organisms per deciliter does *not* exceed the EPA limit.

6. Fifty percent of the values are on either side of the mean. Since 3% + 20% + 21% = 44% while 3% + 20% + 21% + 23% = 67%, the median number of partners for men must be in the 5–10 category. Similarly, the median number of partners for women is in the 2–4 category. (The actual answers from the study are a median of 6 for men and 2 for women.)

## Chapter 2

1. Mean is 147.55, range is 82, and standard deviation is 25.13.

2. a. 17.59, 18.41, 5.94
   b. 19.09, 18.41, 5.94; 20.23, 21.17, 6.83

3. $0.17, $0.19, $0.06

## Chapter 3

1. 86%, 92%, 4%

2. a. $(221 − 206)/35 = 0.43$
      $(216 − 206)/35 = 0.29$
      $(168 − 206)/35 = −1.09$
   b. $206 + 2.69(35) = 300$
      $206 + 0.28(35) = 216$
      $206 − 1.09(35) = 168$

3. Fifth out of 20, 75%, $(14.23−18.87)/5.33 = −0.87$

4. $(10.1 − 7.7)/2.5 = 0.96$

5. $(2319 − 1751)/552 = 1.03$

6. a. $62.5 \pm 2(18.0) = 62.5 \pm 36.0$ or $(26.5, 98.5)$
   b. The index will fall below 26.5 about $5\%/2 = 2.5\%$ of the time.

7. Use Chebyshev's theorem.
   a. $475 \pm 2(55) = (365, 585)$
   b. $(475 - 310)/55 = 3$ and $(640 - 475)/55 = 3$; $1 - 1/3^2 = 88.9\%$.

# Chapter 4

1.

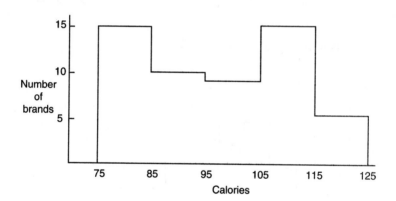

2.

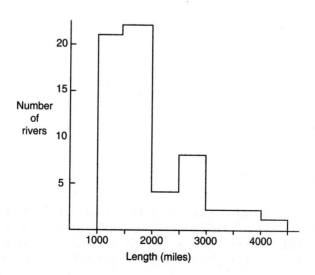

3. First calculate $\mu = 138.7$ and $\sigma = 41.0$.

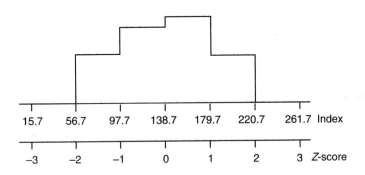

4. First calculate $\mu = 54.83$ and $\sigma = 6.22$.

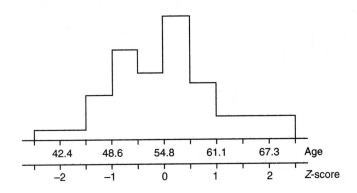

5. a.

```
11 | 0  7
12 | 6  2  7  8  2  7  5  7  8  5  0
13 | 3  7  7
14 | 9  7  8  1  6
15 | 0  9  0  5  2  5  0  4  4
16 | 4  5  9  4  8  2  8  6
17 | 7  6  4  6
18 | 9
19 | 5  3  0  3
20 | 7
21 | 9
22 | 8
```

b. Arranging the set in numerical order gives {110, 117, 120, 122, 122, 125, 125, 126, 127, 127, 127, 128, 128, 133, 137, 137, 141, 146, 147, 148, 149, 150, 150, 150, 152, 154, 154, 155, 155, 159, 162, 164, 164, 165, 166, 168, 168, 169, 174, 176, 176, 177, 189, 190, 193, 193, 195, 207, 219, 228}. The smallest value is 110, the largest 228. The median is 153, the median of the top half 169, the median of the bottom 128.

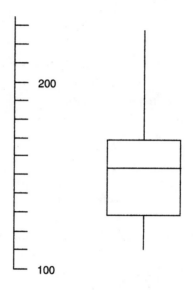

6. a.

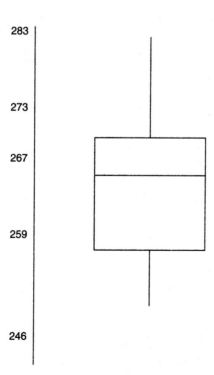

b.

| 28 | 3 | 2 | 3 | | | | | | | | | | |
|----|---|---|---|---|---|---|---|---|---|---|---|---|---|
| 27 | 0 | 1 | 1 | 2 | 2 | 3 | 4 | 4 | 4 | 7 | 7 | 8 | 8 |
| 26 | 0 | 0 | 1 | 2 | 4 | 4 | 5 | 5 | 6 | 7 | 7 | 7 | 7 | 9 |
| 25 | 1 | 5 | 7 | 8 | 8 | 8 | 9 | 9 | 9 | | | | |
| 24 | 6 | 9 | | | | | | | | | | | |

c. $\mu = 266.4$ and $\sigma = 9.0$

d. 75%; 40 of 41, or 97.6%, are in (248.4, 284.4)

e. 68%, 95%, 99%, 70.7%, 97.6%, 100%

7. The median is less than the mean, so the responses are probably skewed to the right; there are a few high guesses with most of the responses on the lower end of the scale.

# Chapter 5

1. Using the multiplicative rule, Michener calculated $13 \times 17 \times 23 \times 37 = 188{,}071$.

2. Boston Chicken advertisers calculated the permutation $P(16,3) = 3360$, when they should have calculated the combination $C(16,3) = 560$.

3. Using the multiplicative rule gives $2^7 = 128$.

# Chapter 6

1. a. $1/4 + 1/7 - (1/4)(1/7) = 5/14$
   b. $1/4$

2. Marilyn assumed *independence* of events (success for surgery $A$ and surgery $B$), which is most likely wrong!

3. If $p = 1/9{,}000{,}000{,}000$, then $P(0 \text{ errors}) = (1 - p)^{20,000,000} = .99778$ and $P(\text{at least 1 error}) = 1 - .99778 = .00222$.

4. $10(.15)^2(.85)^3 = .138$
   $1 - [(.85)^5 + 5(.15)(.85)^4] = .165$
   $(.85)^5 + 5(.15)(.85)^4 + 10(.15)^2(.85)^3 = .973$

5. a. $.5$
   b. $(.5)^8 = .0039$

6. $1 - (.99975)^{100} = .0247$

# Chapter 7

1. $.001(1{,}500{,}000) = 1500$

2. $.03(1200) = 36$

3. $\mu = 0(.6) + 1000(.3) + 10000(.1) = \$1300$
   $\sigma^2 = (0 - 1300)^2(.6) + (1000 - 1300)^2(.3) + (10{,}000 - 1300)^2(.1) = 8{,}610{,}000$ and $\sigma = \$2934$

4. $\mu = \$23,400$ and $\sigma = \$31,350$

5. a. $20,000(.625) = 12,500$
   b. $\mu = 12,500$, $\sigma^2 = 4687.5$, $\sigma = 68.5$

6. $\mu = 25(.73) = 18.25$, $\sigma = 2.22$

## Chapter 8

1. $P(0) = (.45)^3 = .091125$, $P(1) = 3(.55)(.45)^2 = .334125$,
   $P(2) = 3(.55)^2(.45) = .408375$, $P(3) = (.55)^3 = .166375$

2. a. .210, .367, .275, .115, .029, .004, .000, .000
   b. $.029 + .004 = .033$

3. a. .005, .049, .181, .336, .312, .116
   b. $\mu = 5(.065) = 3.25$, $\sigma = 1.067$
   c.

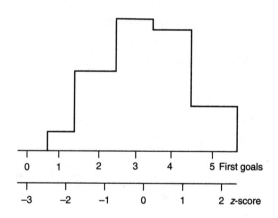

4. a. .017, .087, .195, .260, .228, .137, .057, .016, .003, .000, .000,
   b. $\mu = 3.333$, $\sigma = 1.491$

5. $(.9)^3 + 3(.9)^2(.1) = .972$
   $(.75)^3 + 3(.75)^2(.25) = .844$
   $(.6)^3 + 3(.6)^2(.4) = .648$

## Chapter 9

1. $e^{-1.6} = .202$, $e^{-3.2} = .041$, $e^{-8.0} = .003$

2. $e^{-8} = .003$, $(8^5/5!)e^{-8} = .092$, $(8^8/8!)e^{-8} = .140$

3. $1 - [e^{-4} + 4e^{-4} + (4^2/2!)e^{-4} + (4^3/3!)e^{-4}] = .567$
   $1 - [e^{-2} + 2e^{-2} + (2^2/2!)e^{-2} + (2^3/3!)e^{-2}] = .143$

4. $e^{-2} + 2e^{-2} + (2^2/2!)e^{-2} + (2^3/3!)e^{-2} = .857$
   $e^{-10} + 10e^{-10} + (10^2/2!)e^{-10} + (10^3/3!)e^{-10} = .010$

5. $\mu = 231/336 = 0.631$
$e^{-0.631} = .532$, $0.631e^{-0.631} = .366$, $(0.631^2/2!)e^{-0.631} = .106$, $(0.631^3/3!)$
$e^{-0.631} = .022$, $(0.631^4/4!)e^{-0.631} = .004$
Then $.532 \times 366 = 195$, $.336 \times 366 = 123$, $.106 \times 366 = 39$, $.022 \times 366$
$= 8$, $.004 \times 366 = 1.5$.
(Actual numbers were surprisingly close: 201 days with no deaths, 118 days with one death, 33 days with two deaths, 10 days with three deaths, 3 days with 4 deaths, and 1 day with five deaths.)

# Chapter 10

1. $.5 - .1915 = .3085$, $.5 - .4772 = .0228$

2. $.5 - .2422 = .2578$, $.5 + .4484 = .9484$, $.5 - .1879 = .3121$,
   $.5 + .4982 = .9982$

3. $.2881 + .3849 = .6730$, $.4974 - .2881 = .2093$, $.4772 - .3849 = .0923$

4. $.5 - .4772 = .0228$, $500 - 1.645(100) = 336$, $500$

5. $30 - 2.33(4) = 20.7$, $30 + 1.645(4) = 36.6$

6. $9500 - 1.28(1750) = 7260$, $9500 \pm 1.96(1750) = 6070$ and $12,930$,
   $9500 \pm 2.33(1750) = 5422$ & $13,578$

7. $100 + 0.92\sigma = 120$, $\sigma = 21.7$

8. $\mu + 1.04(2) = 16$, $\mu = 13.92$

9. $\{\mu - 0.67\sigma = 75, \mu + 1.04\sigma = 150\}$, $\mu = 104.39$, $\sigma = 43.86$

10. $c + 2.58(0.4) = 8$, $c = 6.97$

11. $\mu = 29(.81) = 23.49$, $\sigma = 2.11$
    $.5 - .4706 = .0294$, $.1844 - .0047 = .1797$, $.5 - .0047 = .4953$

12. $\mu = 30(.5) = 15$, $\sigma = 2.739$, $.5 - .4495 = .0505$

13. a. $1225(.05)^2(.95)^{48} = .261$, $19600(.05)^3(.95)^{47} = .220$
    $(2.5^2/2!)e^{-2.5} = .256$, $(2.5^3/3!)e^{-2.5} = .214$
    b. $.2422, .2422$

# Chapter 11

1. $4.4 \pm 1.645(.35/\sqrt{49}) = 4.4 \pm 0.082$

2. $28.5 \pm 1.96(1.2/\sqrt{30}) = 28.5 \pm 0.43$

3. a. $64 \pm .74$, $59 \pm .67$
   b. $126 \pm 6.02$, $101 \pm 4.26$
   c. $4029 \pm 100.2$, $4827 \pm 101.4$

4. $\sigma_{\bar{x}} = 1.52/\sqrt{64} = 0.19$, $0.38/0.19 = 2$, $.4772 + .4772 = .9544$

5. $9540 \pm 2.58(1205/\sqrt{30}) = 9540 \pm 567.6$
   $335(9540 \pm 567.6) = \$3,196,000 \pm \$190,000$

6. $1.96(1.1/\sqrt{n}) \le 0.2$, $\sqrt{n} \ge 10.78$, $n \ge 116.2$; choose $n = 117$.

7. $4(50) = 200$

8. $\bar{x} = 22$, $s = 24.8$, $\sigma_{\bar{x}} = 24.8/\sqrt{35} = 4.19$
   $22 \pm 2.33(4.19) = \$22{,}000{,}000 \pm \$9{,}770{,}000$

9. $\bar{x} = 2753.8/49 = 56.2$, $s^2 = 1348.32/48 = 28.09$, $s = 5.3$,
   $\sigma_{\bar{x}} = 5.3/\sqrt{49} = .757$, $56.2 \pm 1.645(.757) = 56.2 \pm 1.25$

10. $\bar{x} = 378/36 = 10.5$, $s^2 = (4123.35 - 378^2/36)/35 = 4.41$, $s = 2.1$,
    $\sigma_{\bar{x}} = 2.1/\sqrt{36} = 0.35$, $10.5 \pm 1.96(0.35) = 10.5 \pm 0.686$

## Chapter 12

1. $\sigma_{\bar{x}} = 0.9/\sqrt{36} = 0.15$, $(3.25 - 3.5)/0.15 = -1.67$,
   $.5 - .4525 = .0475$

2. $\sigma_{\bar{x}} = 4.50/\sqrt{40} = 0.712$, $c = 25 + 1.645(0.712) = 26.17$

3. $c = 12 + 2.05(6/\sqrt{64}) = 13.54$. Since $13.1 < 13.54$, there is *not* sufficient evidence to reject the city spokesperson's claim.

4. a. $\sigma_{\bar{x}} = 3.5/\sqrt{49} = 0.5$, $(16.25 - 15)/0.5 = 2.5$, $p = .5 - .4938 = .0062$.
   b. Since $.0062 < .01$, there *is* sufficient evidence to doubt the 15-minute claim at all three levels.

5. a. $(1.72 - 1.6)/(0.4/\sqrt{64}) = 2.4$, $.5 - .4918 = .0082$, and because this is a two-tailed test, $p = 2(.0082) = .0164$.
   b. Thus the claim *should* be disputed at the 10% and 5% significance levels but *not* at the 1% level.

6. a. $\bar{x} = 109.08/36 = 3.03$, $s^2 = 0.504/35 = 0.0144$, $s = 0.12$, $\sigma_{\bar{x}} = 0.12/\sqrt{36} = 0.02$, $(3.03 - 3)/0.02 = 1.5$, $p = .5 - .4332 = .0668$.
   b. Since $.0668$ is greater than $.01$, $.02$, and $.05$, the manager's claim should *not* be rejected at the 1%, 2%, or 5% significance level.

7. a. $\bar{x} = 2370/30 = 79$, $s^2 = (197{,}287 - 2370^2/30)/29 = 346.79$, $s = 18.62$,
   $\sigma_{\bar{x}} = 18.62/\sqrt{30} = 3.40$, $(79 - 85)/3.4 = -1.76$, $p = .5 - .4608 = .0392$.
   b. Since $.0392 < .05$, there *is* sufficient evidence to reject the agent's claim at both the 10% and 5% levels of significance.

8. $\sigma_{\bar{x}} = 1.2/\sqrt{36} = 0.2$, $(2.5 - 2.6)/0.2 = -0.5$, $\beta = .5 + .1915 = .6915$. $(2.5 - 2.4)/0.2 = 0.5$, $\beta = .5 - .1915 = .3085$.

9. a. $\sigma_{\bar{x}} = 8/\sqrt{100} = 0.8$, $c = 25 - 2.33(0.8) = 23.14$.
   b. For 24, $(23.14 - 24)/0.8 = -1.08$, $\beta = .5 + .3599 = .8599$.
   For 22, $(23.14 - 22)/0.8 = 1.43$, $\beta = .5 - .4236 = .0764$.
   For 20, $(23.14 - 20)/0.8 = 3.925$, $\beta = .0000$.

10. $\sigma_{\bar{x}} = 0.9/\sqrt{36} = 0.15$.
    For $\alpha = .10$, $c = 1.5 + 1.28(0.15) = 1.692$, $(1.692 - 1.8)/0.15 = -0.72$, $\beta = .5 - .2642 = .2358$.
    For $\alpha = .05$, $c = 1.5 + 1.645(0.15) = 1.747$, $(1.747 - 1.8)/0.15 = -0.35$, $\beta = .5 - .1368 = .3632$.

For $\alpha = .01$, $c = 1.5 + 2.33(0.15) = 1.850$, $(1.850 - 1.8)/0.15 = 0.33$,
$\beta = .5 + .1293 = .6293$.

11. a. $\bar{x} = 682.69$, $s = 290.91$, $\sigma_{\bar{x}} = 290.91/\sqrt{80} = 32.52$,
$(682.69 - 750)/32.52 = -2.07$, $p = .5 - .4808 = .0192$

b. There *is* sufficient evidence that the average radiation in North American sites is less than 750 BTU at the 5% significance level, but *not* at the 1% level.

## Chapter 13

1.
$$\sigma_{\bar{x}-\bar{y}} = \sqrt{\frac{(19.1)^2}{347} + \frac{(19.9)^2}{561}} = 1.326$$

$(93.5 - 84.5) \pm 1.96(1.326) = 9.0 \pm 2.60$

2.
$$\sigma_{\bar{x}-\bar{y}} = \sqrt{\frac{(6.3)^2}{274} + \frac{(6.3)^2}{90}} = 0.765$$

$(33.0 - 28.6) \pm 1.645(0.765) = 4.4 \pm 1.26$

3.
$$\bar{x}_1 = \frac{208,000}{400} = 520 \qquad \bar{x}_2 = \frac{255,000}{500} = 510$$

$$s_1 = \sqrt{\frac{760,000}{400 - 1}} = 43.64 \qquad s_2 = \sqrt{\frac{1,050,000}{500 - 1}} = 45.87$$

$$\sigma_{\bar{x}-\bar{y}} = \sqrt{\frac{(43.64)^2}{400} + \frac{(45.87)^2}{500}} = 2.995$$

$(520 - 510) \pm 2.33(2.995) = 10 \pm 6.98$

4.
$$\sigma_{\bar{x}-\bar{y}} = \sqrt{\frac{(4.5)^2}{n} + \frac{(4.5)^2}{n}} = \frac{6.364}{\sqrt{n}}$$

$1.645(6.364/\sqrt{n}) \leq 1$, $\sqrt{n} \geq 10.47$, $n \geq 109.6$, so choose $n = 110$.

5.
$$\sigma_{\bar{x}-\bar{y}} = \sqrt{\frac{(329)^2}{231} + \frac{(264)^2}{221}} = 28.00$$

$c = 0 - 2.33(28.00) = -65.2$, $6176 - 8164 = -1988$. The difference *is* significant, and there *is* sufficient evidence at the 1% significance level to say that angioplasties cost less than bypass surgeries.

6.
$$\sigma_{\bar{x}-\bar{y}} = \sqrt{\frac{(10.2)^2}{53} + \frac{(11.2)^2}{53}} = 2.081$$

$c = 0 - 1.645(2.081) = -3.42$, $83.5 - 88.1 = -4.6$.
There *is* sufficient evidence at the 5% significance level to say that average IQ improves after treatment for lead poisoning. (See also the discussion on paired differences in the Study Questions.)

7.
$$\bar{x}_1 = \frac{1,300}{500} = 2.6 \qquad \bar{x}_2 = \frac{896}{350} = 2.56$$

$$s_1 = \sqrt{\frac{3,460 - \dfrac{1300^2}{500}}{500 - 1}} = .400 \qquad s_2 = \sqrt{\frac{2,360 - \dfrac{896^2}{350}}{350 - 1}} = .436$$

$$\sigma_{\bar{x}-\bar{y}} = \sqrt{\frac{(.400)^2}{500} + \frac{(.436)^2}{350}} = .0294$$

$c = 0 + 1.28(.0294) = .0376$, $2.6 - 2.56 = .04$
There *is* sufficient evidence at the 10% significance level to say that liberal arts colleges give higher grades.

8. Standard deviations from the two samples are needed.

# Chapter 14

1. $df = 15$, $7.4 \pm 2.947(1.3/\sqrt{16}) = 7.4 \pm .96$

2. $\bar{x} = 35.7/17 = 2.1$, $s^2 = 0.096/16 = 0.006$, $s = 0.0775$, $df = 16$, $2.1 \pm 1.337(0.0775/\sqrt{17}) = 2.1 \pm 0.025$

3. $\bar{x} = 4.048$, $s = 2.765$, $df = 20$, $4.048 \pm 1.725(2.765/\sqrt{21}) = 4.048 \pm 1.041$

4. a. $7.5 \pm 2.262(4.085/\sqrt{9}) = 7.5 \pm 3.08$
   b. $19.11 \pm 2.262(6.791/\sqrt{9}) = 19.11 \pm 5.12$
   c. $11.61 \pm 2.262(4.891/\sqrt{9}) = 11.61 \pm 3.69$

5. $c = 1250 - 1.796(616/\sqrt{12}) = 930.6$
   Since $934 > 930.6$, there is *not* sufficient evidence at the 5% significance level to dispute the IRS claim.

6. Critical values are $2000 \pm 2.093(600/\sqrt{20}) = 1719.2$ and $2280.8$. Since 2250 is inside this range, there is *not* sufficient evidence at the 5% significance level to dispute the auditor's remarks.

7. $\bar{x} = 50.4/18 = 2.8$, $s^2 = (163.8 - 50.4^2/18)/17 = 1.334$, $s = 1.155$, $c = 2 + 2.567(1.155/\sqrt{18}) = 2.70$.
   Since $2.8 > 2.70$, there *is* sufficient evidence at the 1% significance level to dispute the 2-hour claim.

8. $\bar{x} = 5.67$, $s = 0.763$, $c = 5 + 2.015(0.763/\sqrt{6}) = 5.63$.
   Since $5.67 > 5.63$, there *is* sufficient evidence at the 5% significance level to dispute the company scientist's claim.

9. $\bar{x} = 24.581$, $s = 0.7793$, $c = 25 - 2.485(0.7793/\sqrt{26}) = 24.620$. Since 24.581 $< 24.620$, there *is* sufficient evidence at the 1% significance level to dispute the null hypothesis of 25.

10. $\bar{x} = 301.3$, $s = 2.21$, critical values are $300 \pm 3.143(2.21/\sqrt{7}) = 297.37$ and $302.63$. Since 301.3 is inside this range, there is *not* sufficient evidence at the 2% significance level for the FDA investigator to reject the label.

11.
$$\sigma_{\bar{x}-\bar{y}} = \sqrt{\frac{(6-1)(1.4)^2 + (6-1)(1.7)^2}{6+6-2}} \sqrt{\frac{1}{6}+\frac{1}{6}} = 0.899$$

$(75 - 71) \pm 3.169(0.899) = 4 \pm 2.85$

12.
$$\sigma_{\bar{x}-\bar{y}} = \sqrt{\frac{(8-1)(2.3)^2 + (5-1)(2.8)^2}{8+5-2}} \sqrt{\frac{1}{8}+\frac{1}{5}} = 1.421$$

$(8.1 - 9.8) \pm 2.201(1.421) = -1.7 \pm 3.13$

13.
$$n_1 = 5 \qquad n_2 = 7$$

$$\bar{x}_1 = 61 \qquad \bar{x}_2 = 78$$

$$s_1 = 3.54 \qquad s_2 = 10.23$$

$$\sigma_{\bar{x}-\bar{y}} = \sqrt{\frac{(5-1)(3.54)^2 + (7-1)(10.23)^2}{5+7-2}} \sqrt{\frac{1}{5}+\frac{1}{7}} = 4.822$$

$(61 - 78) \pm 1.812(4.822) = -17 \pm 8.74$

14.
$$\sigma_{\bar{x}-\bar{y}} = \sqrt{\frac{(10-1)(0.3)^2 + (10-1)(0.2)^2}{10+10-2}} \sqrt{\frac{1}{10}+\frac{1}{10}} = 0.114$$

$c = 0 - 1.734(0.114) = -0.198$.
Since $4 - 4.8 = -0.8 < -0.198$, there *is* sufficient evidence at the 5% significance level that the new diet improves weight gain in laboratory mice.

15.
$$n_1 = 8 \qquad n_2 = 6$$

$$\bar{x}_1 = 70 \qquad \bar{x}_2 = 67$$

$$s_1 = 8.32 \qquad s_2 = 9.98$$

$$\sigma_{\bar{x}-\bar{y}} = \sqrt{\frac{(8-1)(8.32)^2 + (6-1)(9.98)^2}{8+6-2}} \sqrt{\frac{1}{8}+\frac{1}{6}} = 4.89$$

Critical values are $0 \pm 1.356(4.89) = \pm 6.63$.
Since $70 - 67 = 3$ is inside the range, the observed difference in typing speeds is *not* significant with $\alpha = .20$.

## Chapter 15

1. $\bar{x}_1 = 3.75$, $s_1 = 0.5$, $\bar{x}_2 = 1.25$, $s_2 = 0.957$, $\bar{x}_3 = 2.75$, $s_3 = 1.258$
$\mu = (3.75 + 1.25 + 2.75)/3 = 2.583$
$\sigma_{\bar{x}}^2 = [(3.75 - 2.583)^2 + (1.25 - 2.583)^2 + (2.75 - 2.583)^2]/(3-1) = 1.583$
$\sigma_{within}^2 = (0.5^2 + 0.957^2 + 1.258^2)/3 = 0.916$
$\sigma_{between}^2 = 4(1.583) = 6.332$
$F = 6.332/0.916 = 6.91$
With $df_{numerator} = 3 - 1 = 2$, $df_{denominator} = 3(4-1) = 9$, and $\alpha = .05$, the critical $F$-value is $F_{critical} = 4.26$. Since $6.91 > 4.26$,

there *is* sufficient evidence that the three honeys do not have the same effectiveness.

2. $\mu = (23.0 + 12.9 + 8.6)/3 = 14.83$
$\sigma_{\bar{x}}^2 = [(23.0 - 14.83)^2 + (12.9 - 14.83)^2 + (8.6 - 14.83)^2]/(3 - 1) = 54.64$
$\sigma_{within}^2 = (11.2^2 + 8.8^2 + 6.4^2)/3 = 81.28$
$\sigma_{between}^2 = 25(54.64) = 1366$
$F = 1366/81.28 = 16.81$
With $df_{numerator} = 3 - 1 = 2$, $df_{denominator} = 3(25 - 1) = 72$, and $\alpha = .01$, the critical $F$-value is $F_{critical} = 4.94$. Since $16.81 > 4.94$, there *is* sufficient evidence that the three treatments produce different results.

3. $\bar{x}_1 = 3.73$, $\bar{x}_2 = 4.27$, $\bar{x}_3 = 3.97$, $\bar{x}_4 = 3.77$, $\bar{x}_5 = 4.47$
$s_1 = 0.904$, $s_2 = 0.594$, $s_3 = 0.769$, $s_4 = 0.704$, $s_5 = 0.896$
$\mu = (3.73 + 4.27 + 3.97 + 3.77 + 4.47)/5 = 4.042$
$\sigma_{\bar{x}}^2 = [(3.73-4.042)^2 + (4.27-4.042)^2 + (3.97-4.042)^2 + (3.77-4.042)^2 + (4.47-4.042)^2]/(5-1) = 0.103$
$\sigma_{within}^2 = (0.904^2 + 0.594^2 + 0.769^2 + 0.704^2 + 0.896^2)/5 = 0.612$
$\sigma_{between}^2 = 15(0.103) = 1.545$
$F = 1.545/0.612 = 2.52$
With $df_{numerator} = 5 - 1 = 4$, $df_{denominator} = 5(15 - 1) = 70$, and $\alpha = .01$, the critical $F$-value is $F_{critical} = 3.62$. Since $2.52 < 3.62$, there is *not* sufficient evidence at the 1% significance level to conclude that some franchises prepare orders quicker than others.

# Chapter 16

1.
$$\sigma_{\bar{p}} = \sqrt{\frac{(.39)(.61)}{500}} = .0218$$

$.39 \pm 1.645(.0218) = .39 \pm .036$

2.
$$\sigma_{\bar{p}} = \sqrt{\frac{(.75)(.25)}{1000}} = .0137$$

$.75 \pm 1.96(.0137) = .75 \pm .027$

3. a. $\bar{p} = 26/130 = .2$

$$\sigma_{\bar{p}} = \sqrt{\frac{(.2)(.8)}{130}} = .0351$$

$.2 \pm 2.58(.0351) = .2 \pm .091$
b. $13,000(.2 \pm .091) = 2,600 \pm 1,180$

4. a. $.5 \pm 2.33(.0236) = .5 \pm .055$
b. $.32 \pm 1.88(.0342) = .32 \pm .064$
c. $.31 \pm 1.28(.0661) = .31 \pm .085$

5.
$$\sigma_{\bar{p}} = \sqrt{\frac{(.17)(.83)}{1703}} = .0091$$

$z(.0091) = .02$, $z = 2.20$, $2(.4861) = 97.2\%$

6.
$$\sigma_{\bar{p}} = \sqrt{\frac{(.75)(.25)}{60}} = .0559$$

$c = .75 - 1.645(.0559) = .658 = 65.8\%$

7. $1.645(.5/\sqrt{n}) \le .04$, $\sqrt{n} \ge 20.563$, $n \ge 422.8$, so choose $n = 423$.

8. a. 95%

   b. $1.96(.5/\sqrt{n}) \le .03$, $\sqrt{n} \ge 32.67$, $n \ge 1067.1$, so the pollsters should have obtained a sample size of at least 1068. (They actually interviewed 1148 people.)

## Chapter 17

1.
$$\sigma_{\bar{p}} = \sqrt{\frac{(.66)(.34)}{5,644}} = .0063$$

$c = .66 - 2.33(.0063) = .645$.
Since $.35 < .645$ (in fact, a lot less), there *is* sufficient evidence to say that the rate of sharing needles has dramatically dropped.

2.
$$\sigma_{\bar{p}} = \sqrt{\frac{(.0071)(.9929)}{7,782}} = .00095$$

Critical values are $.0071 \pm 1.96(.00095) = .00524$ and $.00896$. Since $\bar{p} = 29/7782 = .00373$ is not in this range, there *is* sufficient evidence to dispute the 0.71% claim.

3. a.
$$\sigma_{\bar{p}} = \sqrt{\frac{(.10)(.90)}{3,432}} = .0051$$

$(.021 - .10)/.0051 = -15.5$, and the *p*-value is .0000.

   b. There *is* sufficient evidence to reject the Kinsey study's claim.

4. a.
$$\sigma_{\bar{p}} = \sqrt{\frac{(.40)(.60)}{27}} = .0943$$

$\bar{p} = 16/27 = .593$, $(.593 - .40)/.0943 = 2.047$ and the *p*-value is $.5 - .4798 = .0202$.

   b. Since $.0202 < .025$, there is sufficient evidence at the 2.5% significance level to claim that more than 40% of coffee drinkers show signs of dependence.

5. a.
$$\sigma_{\bar{p}} = \sqrt{\frac{(.13)(.87)}{200}} = .0238$$

$(.10 - .13)/.0238 = -1.26$ and $\alpha = .5 - .3962 = .1038$.

   b. For .12, $(.10-.12)/.0238 = -0.84$ and $\beta = .5 + .2995 = .7995$.
   For .09, $(.10 - .09)/.0238 = 0.42$ and $\beta = .5 - .1628 = .3372$.

6.
$$\sigma_{\bar{p}} = \sqrt{\frac{(.40)(.60)}{500}} = .0219$$

$c = .4 + 1.28(.0219) = .428$
For 42%, $(.428 - .42)/.0219 = 0.37$ and $\beta = .5 + .1443 = .6443$.
For 44%, $(.428 - .44)/.0219 = -0.55$ and $\beta = .5 - .2088 = .2912$.

7.
$$\sigma_{\bar{p}} = \sqrt{\frac{(.82)(.18)}{230}} = .0253$$

$c = .82 - 1.645(.0253) = .778$
For 81%, $(.778 - .81)/.0253 = -1.26$ and $\beta = .5 + .3962 = .8962$
For 80%, $(.778 - .80)/.0253 = -0.87$ and $\beta = .5 + .3078 = .8078$
For 79%, $(.778 - .79)/.0253 = -0.47$ and $\beta = .5 + .1808 = .6808$
For 78%, $(.778 - .78)/.0253 = -0.08$ and $\beta = .5 + .0319 = .5319$
For 77%, $(.778 - .77)/.0253 = \phantom{-}0.32$ and $\beta = .5 - .1255 = .3745$
For 76%, $(.778 - .76)/.0253 = \phantom{-}0.71$ and $\beta = .5 - .2611 = .2389$
For 75%, $(.778 - .75)/.0253 = \phantom{-}1.11$ and $\beta = .5 - .3665 = .1335$
For 74%, $(.778 - .74)/.0253 = \phantom{-}1.50$ and $\beta = .5 - .4332 = .0668$
For 73%, $(.778 - .73)/.0253 = \phantom{-}1.90$ and $\beta = .5 - .4713 = .0287$

# Chapter 18

1.
$$n_1 = 361 \qquad\qquad n_2 = 86$$

$$\bar{p}_1 = \frac{210}{361} = .582 \qquad \bar{p}_2 = \frac{34}{86} = .395$$

$$\sigma_d = \sqrt{\frac{(.582)(.418)}{361} + \frac{(.395)(.605)}{86}} = .0588$$

$(.582 - .395) \pm 1.96(.0588) \qquad\qquad = .187 \pm .115$

2.
$$n_1 = 300 \qquad n_2 = 400$$

$$\bar{p}_1 = .65 \qquad \bar{p}_2 = .48$$

$$\sigma_d = \sqrt{\frac{(.65)(.35)}{300} + \frac{(.48)(.52)}{400}} = .0372$$

$(.65 - .48) \pm 2.58(.0372) \qquad\qquad = .17 \pm .096$

3.
$$n_1 = 94 \qquad\qquad n_2 = 80$$

$$\bar{p}_1 = \frac{23}{94} = .245 \qquad \bar{p}_2 = \frac{34}{80} = .425$$

$$\sigma_d = \sqrt{\frac{(.245)(.755)}{94} + \frac{(.425)(.575)}{80}} = .0709$$

$(.245 - .425) \pm 1.645(.0709) \qquad\qquad = -.18 \pm .117$

4.
$$\bar{p}_1 = \frac{50}{228} = .219 \qquad \bar{p}_2 = \frac{2}{64} = .031$$

$$\bar{p} = \frac{50 + 2}{222 + 64} = .182$$

$$\sigma_d = \sqrt{(.182)(.818)\left(\frac{1}{228} + \frac{1}{64}\right)} = .0546$$

$c = 0 + 1.645(.0546) = .107$.

Since $.219 - .031 = .188 > .107$, there *is* sufficient evidence at the 5% significance level to say that a higher proportion of blacks than whites receive the death penalty.

5.
$$\bar{p}_1 = \frac{48}{53} = .906 \qquad \bar{p}_2 = \frac{251}{285} = .881$$

$$\bar{p} = \frac{48 + 251}{53 + 285} = .885$$

$$\sigma_d = \sqrt{(.885)(.115)\left(\frac{1}{53} + \frac{1}{285}\right)} = .0477$$

Critical values are $0 \pm 1.96(.0477) = \pm .0935$.

Since $.906 - .881 = .025$ is within this range, there is *not* sufficient evidence at the 5% significance level to say that the probability Bird will make a second free throw depends on whether he makes the first.

6.
$$.083(240) = 20 \qquad .255(240) = 61$$

$$\bar{p} = \frac{20 + 61}{240 + 240} = .169$$

$$\sigma_d = \sqrt{(.169)(.831)\left(\frac{1}{240} + \frac{1}{240}\right)} = .0342$$

$c = 0 - 2.33(.0342) = -.0797$.

Since $.083 - .255 = -.172 < -.0797$, there *is* sufficient evidence at the 1% significance level to say that AZT helps prevent the HIV virus from being passed from mother to child.

7.
$$.92(875) = 805 \qquad .93(875) = 814$$

$$\bar{p} = \frac{805 + 814}{875 + 875} = .925$$

$$\sigma_d = \sqrt{(.925)(.075)\left(\frac{1}{875} + \frac{1}{875}\right)} = .0126$$

$(.92 - .93)/.0126 = -0.79$ and the *p*-value of this two-tailed test is $2(.5 - .2852) = .4296$.

With such a high *p*-value, the difference in survival rates is *not* statistically significant.

8. a.
$$\bar{p}_1 = \frac{2599}{4710} = .552 \qquad \bar{p}_2 = \frac{1084}{2979} = .364$$

$$\sigma_d = \sqrt{\frac{(.552)(.448)}{4710} + \frac{(.364)(.636)}{2979}} = .0114$$

$(.552 - .364) \pm 2.05(.0114) \qquad\qquad = .188 \pm .0234$

b.
$$\bar{p}_1 = \frac{551}{602} = .915 \qquad \bar{p}_2 = \frac{383}{856} = .447$$

$$\sigma_d = \sqrt{\frac{(.915)(.085)}{602} + \frac{(.447)(.553)}{856}} = .0204$$

$(.915 - .447) \pm 2.33(.0204) \qquad\qquad = .468 \pm .0475$

c.
$$\bar{p}_1 = \frac{383}{4710} = .081 \qquad \bar{p}_2 = \frac{473}{2979} = .159$$

$$\bar{p} = \frac{383 + 473}{4710 + 2979} = .111$$

$$\sigma_d = \sqrt{(.111)(.889)\left(\frac{1}{4710} + \frac{1}{2979}\right)} = .0074$$

$c = 0 - 2.58(.0074) = -.0190.$
Since $.081 - .159 = -.078 < -.0190$, there *is* sufficient evidence that a greater proportion of women than men earn under $10,000.

d.
$$\bar{p}_1 = \frac{3776}{4710} = .802 \qquad \bar{p}_2 = \frac{2455}{2979} = .824$$

$$\bar{p} = \frac{3776 + 2455}{4710 + 2979} = .810$$

$$\sigma_d = \sqrt{(.810)(.190)\left(\frac{1}{4710} + \frac{1}{2979}\right)} = .0092$$

Critical values are $0 \pm 2.24(.0092) = \pm .021$.
Since $.802 - .824 = -.022$ is outside this range, the difference between the proportions of men and women earning between $10,000 and $50,000 *is* significant at the 2.5% significance level.

9. $1.645(.5\sqrt{2}/\sqrt{n}) \le .03$, $\sqrt{n} \ge 38.77$, $n \ge 1503.3$, so the researcher should choose a sample size of at least 1504.

# Chapter 19

1. $.20(625) = 125$, $.25(625) = 156.25$, $.35(625) = 218.75$

$$\chi^2 = \frac{(110 - 125)^2}{125} + \frac{(140 - 156.25)^2}{156.25} + \frac{(250 - 218.75)^2}{218.75}$$

$$+ \frac{(125 - 125)^2}{125} = 7.954$$

With $df = 4 - 1 = 3$ and $\alpha = .05$, the critical $\chi^2$ value is 7.815. Since 7.954 > 7.815, there *is* sufficient evidence at the 5% significance level to say that the percentages have changed.

2. We have that $2 + 3 + 3 + 4 + 6 = 18$ and

$$\frac{2}{18}(6000) = 667 \qquad \frac{3}{18}(6000) = 1000$$

$$\frac{4}{18}(6000) = 1333 \qquad \frac{6}{18}(6000) = 2000$$

$$\chi^2 = \frac{(720 - 667)^2}{667} + \frac{(970 - 1000)^2}{1000} + \frac{(1013 - 1000)^2}{1000}$$

$$+ \frac{(1380 - 1333)^2}{1333} + \frac{(1917 - 2000)^2}{2000} = 10.38$$

With $df = 5 - 1 = 4$ and $\alpha = .05$ the critical $\chi^2$ value is 9.448; with $\alpha = .025$ the critical $\chi^2$ value is 11.14. Thus there *is* sufficient evidence to reject the superintendent's claim at the 5% significance level but not at the 2.5% level.

3. $(1372 + 1578 + 1686)/3 = 1545$

$$\chi^2 = \frac{(1372 - 1545)^2}{1545} + \frac{(1578 - 1545)^2}{1545} + \frac{(1686 - 1545)^2}{1545} = 32.94$$

With $df = 3 - 1 = 2$ and $\alpha = .05$, the critical $\chi^2$ value is 5.991. Since 32.94 > 5.991, there *is* sufficient evidence at the 5% significance level to say that the number of accidents on each shift are not the same.

4. $P(\leq 2) = (.2)^5 + 5(.8)(.2)^4 + 10(.8)^2(.2)^3 = .05792$, $P(3) = 10(.8)^3(.2)^2 = .2048$, $P(4) = 5(.8)^4(.2) = .4096$, $P(5) = (.8)^5 = .32768$, $.05792(200) = 11.6$, $.2048(200) = 41.0$, $.4096(200) = 81.9$, $.32768(200) = 65.5$.

$$\chi^2 = \frac{(5 - 11.6)^2}{11.6} + \frac{(34 - 41.0)^2}{41.0} + \frac{(90 - 81.9)^2}{81.9} + \frac{(71 - 65.5)^2}{65.5} = 6.213$$

With $df = 4 - 1 = 3$ and $\alpha = .05$, the critical $\chi^2$ value is 7.815. Since 6.213 < 7.815 we conclude that the observed data do *not* differ significantly from what would be expected for a binomial distribution with $\pi = .8$.

5. $245(0) + 170(1) + 50(2) + 35(3) = 375$, $\bar{p} = 375/[3(500)] = .25$. $(.75)^3(500) = 211$, $3(.25)(.75)^2(500) = 211$, $3(.25)^2(.75)(500) = 70$, $(.25)^3(500) = 8$.

$$\chi^2 = \frac{(245 - 211)^2}{211} + \frac{(170 - 211)^2}{211} + \frac{(50 - 70)^2}{70} + \frac{(35 - 8)^2}{8} = 110.3$$

With $df = 4 - 2 = 2$ and $\alpha = .01$, the critical $\chi^2$ value is 9.21. Since 110.3 > 9.21 there *is* sufficient evidence that the distribution is *not* binomial.

6.
$$P(0) = e^{-2.4} = .091 \qquad P(1) = 2.4e^{-2.4} = .218$$

$$P(2) = \frac{(2.4)^2}{2} e^{-2.4} = .261 \qquad P(3) = \frac{(2.4)^3}{3!} e^{-2.4} = .209$$

$$P(4) = \frac{(2.4)^4}{4!} e^{-2.4} = .125$$

$$P(5 \text{ or more}) = 1 - (.091+.218+.261+.209+.125) = .096$$

$.091(83) = 7.6, .218(83) = 18.1, .261(83) = 21.7,$
$.209(83) = 17.3, .125(83) = 10.4, .096(83) = 8.0.$

$$\chi^2 = \frac{(3-7.6)^2}{7.6} + \frac{(15-18.1)^2}{18.1} + \frac{(28-21.7)^2}{21.7} + \frac{(25-17.3)^2}{17.3}$$
$$+ \frac{(6-10.4)^2}{10.4} + \frac{(4-8.0)^2}{8.0} = 51.00$$

With $df = 6 - 1 = 5$ and $\alpha = .05$, the critical $\chi^2$ value is 11.07. Since $51.00 > 11.07$, there *is* sufficient evidence that the distribution is *not* Poisson with $\mu = 2.4$.

7. $5(0)+16(1)+17(2)+8(3)+4(4) = 90$, $90/50 = 1.8$.

$$e^{-1.8}(50) = 8.3 \qquad 1.8 e^{-1.8}(50) = 14.9 \qquad \frac{(1.8)^2}{2} e^{-1.8}(50) = 13.4$$

$$\frac{(1.8)^3}{3!} e^{-1.8}(50) = 8.0 \qquad \frac{(1.8)^4}{4!} e^{-1.8}(50) = 3.6$$

$$\chi^2 = \frac{(5-8.3)^2}{8.3} + \frac{(16-14.9)^2}{14.9} + \frac{(17-13.4)^2}{13.4} + \frac{(8-8.0)^2}{8.0}$$
$$+ \frac{(4-3.6)^2}{3.6} = 2.405$$

With $df = 5 - 2 = 3$ and $\alpha = .10$, the critical $\chi^2$ value is 9.236. Since $2.405 < 9.236$, the observed data do *not* differ significantly from what would be expected for a Poisson distribution.

8. $.0668(100) = 6.68, .2417(100) = 24.17, .3830(100) = 38.30$

$$\chi^2 = \frac{(10-6.68)^2}{6.68} + \frac{(32-24.17)^2}{24.17} + \frac{(33-38.3)^2}{38.3} + \frac{(15-24.17)^2}{24.17}$$
$$+ \frac{(10-6.68)^2}{6.68} = 10.049$$

With $df = 5 - 1 = 4$ and $\alpha = .025$, the critical $\chi^2$-value is 11.14. Since $10.049 < 11.14$, the observed data do *not* differ significantly from what would be expected for a normal distribution with $\mu = 55$ and $\sigma = 10$.

## Chapter 20

1.

| | Vertex baldness | No vertex baldness | |
|---|---|---|---|
| Heart attacks | 214 | 451 | 665 |
| Other admissions | 175 | 597 | 772 |
| | 389 | 1048 | |

| Expected | |
|---|---|
| 180 | 485 |
| 209 | 563 |

$(665)(389)/1437 = 180$, $(665)(1048)/1437 = 485$,
$(772)(389)/1437 = 209$, $(772)(1048)/1437 = 563$

$$\chi^2 = \frac{(214 - 180)^2}{180} + \frac{(451 - 485)^2}{485} + \frac{(175 - 209)^2}{209} + \frac{(597 - 563)^2}{563}$$
$$= 16.39$$

With $df = (2 - 1)(2 - 1) = 1$ and $\alpha = .05$, the critical $\chi^2$ value is 3.841. Since $16.39 > 3.841$, there *is* sufficient evidence at the 5% significance level to say that a relationship exists between heart attacks and vertex baldness.

2.

| | Aspirin | Placebo | |
|---|---|---|---|
| Fatal attack | 11.5 | 11.5 | 23 |
| Nonfatal attack | 135 | 135 | 270 |
| No heart attack | 10,890.5 | 10,887.5 | 21,778 |
| | 11,037 | 11,034 | |

$$\chi^2 = \frac{(5 - 11.5)^2}{11.5} + \frac{(18 - 11.5)^2}{11.5} + \frac{(99 - 135)^2}{135} + \frac{(171 - 135)^2}{135}$$
$$+ \frac{(10,933 - 10,890.5)^2}{10,890.5} + \frac{(10,845 - 10,887.5)^2}{10,887.5} = 26.88$$

With $df = (3 - 1)(2 - 1) = 2$ and $\alpha = .05$, the critical $\chi^2$-value is 5.991. Since $26.88 > 5.991$, there *is* sufficient evidence at the 5% significance level to say that a relationship exists between taking aspirins and the risk of heart attacks.

3.

| | Low Cholesterol | Medium Cholesterol | High Cholesterol | |
|---|---|---|---|---|
| Nonfatal heart attacks | 27.4 | 21.1 | 15.5 | 64 |
| Fatal heart attacks | 20.6 | 15.9 | 11.5 | 48 |
| | 48 | 37 | 27 | |

$$\chi^2 = \frac{(29 - 27.4)^2}{27.4} + \frac{(17 - 21.1)^2}{21.1} + \frac{(18 - 15.5)^2}{15.5} + \frac{(19 - 20.6)^2}{20.6}$$

$$+ \frac{(20 - 15.9)^2}{15.9} + \frac{(9 - 11.5)^2}{11.5} = 3.018$$

With $df = (2 - 1)(3 - 1) = 2$ and $\alpha = .10$, the critical $\chi^2$-value is 4.605. Since $3.018 < 4.605$ there is *not* sufficient evidence at the 10% significance level of a relationship between cholesterol level and risk of heart disease for men over 70 years old.

4.
$$\chi^2 = \frac{(57 - 71.8)^2}{71.8} + \frac{(34 - 36.4)^2}{36.4} + \frac{(24 - 17.7)^2}{17.7} + \frac{(17 - 6.1)^2}{6.1}$$

$$+ \frac{(548 - 533.2)^2}{533.2} + \frac{(273 - 270.6)^2}{270.6} + \frac{(125 - 131.3)^2}{131.3}$$

$$+ \frac{(34 - 44.9)^2}{44.9} = 28.31$$

With $df = (2 - 1)(4 - 1) = 3$ and $\alpha = .01$, the critical $\chi^2$-value is 13.28. Since $28.31 > 13.28$, there *is* sufficient evidence at the 1% significance level of a relationship between amount of fecal material in the water and reported gastroenteritis illness.

5. Since $4.18 < 9.448$ there is *not* sufficient evidence at the 5% significance level of a relationship between hair loss pattern and body mass index.

6. Since $87.6 > 19.02$ there *is* sufficient evidence at the 2.5% significance level of a relationship between fitness level and smoking habits.

# Chapter 21

1. Slope is $20/10 = 2$, that is, a woman's risk of ovarian cancer rises 2% for every gram of fat consumed per day.

2. a. $y = 0.0227x + 707$ where $x$ represents dollars spent per student and $y$ represents the average SAT total scores.
   b. Each additional $1000 spent per student corresponds to a rise of 22.7 in the average SAT total scores.
   c. $0.0227(10,000) + 707 = 934$
   d. $0.0227x + 707 = 1000$ gives $x = \$12,907$.

3. $\Sigma x = 13{,}923$, $\Sigma y = 108.1$, $\Sigma x^2 = 27{,}692{,}875$, $\Sigma xy = 215{,}050.4$,
   $\overline{x} = 1{,}989$, $\overline{y} = 15.44$
   a. $y = 1.4107x - 2790.47$
   b. The average monthly rate for basic television cable service has increased by about $1.41 per year.
   c. $1.4107(1993) - 2790.47 = \$21.06$
   d. $1.4107x - 2790.47 = 50$ gives 2014 as the year when the rate will reach $50.00.

4. $\Sigma x = 9835$, $\Sigma y = 1267.2$, $\Sigma x^2 = 19{,}346{,}025$, $\Sigma xy = 2{,}490{,}479$,
   $\overline{x} = 1967$, $\overline{y} = 253.44$
   a. $y = -3.627x + 7386.87$
   b. The heart disease rate per 100,000 people has been dropping about 3.627 per year.
   c. $-3.627(1983) + 7386.87 = 194.5$
   d. $-3.627x + 7386.87 = 100$ gives $x = 2009$ as the year when the death rate will drop to 100.

5. $\Sigma x = 294$, $\Sigma y = 52.5$, $\Sigma x^2 = 16{,}812$, $\Sigma xy = 3{,}006$, $\overline{x} = 42$,
   $\overline{y} = 7.5$, $m = [3{,}006 - 7(42)(7.5)]/[16{,}812 - 7(42)^2] = 0.1794$
   a. $y = 0.1794x - 0.03629$
   b. The LIF intensity rises 0.1794 for each unit mtorr increase in the ozone partial pressure.
   c. $0.1794(37) - 0.03629 = 6.6$
   d. $0.1794x - 0.03629 = 18$ gives $x = 100.5$

# Chapter 22

1. a. $y = -2.4348x + 44.256$
   b. $r = -.7206$ and $.7206 > .632$, so there is evidence of correlation.

2. a. $y = 1.926x - 3593.3$
   b. $r = .989$ (Critical $r$ is .575)
   c. 452.1 million
   d. Population is increasing 1.926 million per year.
   e. $300 = 1.926x - 3593.3$ gives $x = 2021$ as the year when the population will reach 300 million.

3. b. $y = 80.39\ e^{0.01298x}$, $r = .996$
   c. 16,288 million
   d. $300 = 80.39\ e^{0.01298x}$ gives $x = 101$, or 2001 as the year when the population will reach 300 million.

4. a. $y = 289.32x - 562{,}474$
   b. $r = .881$ (Critical $r = .576$)
   c. $30,624
   d. Personal income is growing $289.32 per year.
   e. $25{,}000 = 289.32x - 562{,}474$ gives $x = 2031$ as the year when personal income will reach $25,000.

5. b.  $y = 485.34\ e^{0.06534x}$, $r = .993$

c. When $x = 115$, $y = \$59,583$.

$25,000 = 485.34\ e^{0.06534x}$ gives $x = 60$, or 1995 as the year when personal income will reach $\$25,000$.

# GLOSSARY

**α-risk.** The probability of committing a Type I error.

**Average deviation.** Average of the absolute differences of each score from the mean.

**Bar graph.** A visual representation of data in which frequencies of different results are indicated by the heights of bars representing these results.

**β-risk.** The probability of committing a Type II error.

**Bimodal.** When two scores have equal frequency, and this frequency is higher than any other.

**Binomial probabilities.** Probabilities resulting from applications in which a two-outcome situation is repeated some number of times, and the probability of each of the two outcomes remains the same for each repetition.

**Binomial random variable.** One whose values are the number of "successes" in some binomial probability distribution.

**Box and whisker display.** A visual representation of dispersion which shows the largest value, the smallest value, the median, the median of the top half of the set, and the median of the bottom half of the set.

**Central Limit Theorem.** Pick $n$ sufficiently large (at least 30), take all samples of size $n$, and compute the mean of each of these samples. Then the set of these sample means will be approximately *normally* distributed.

**Chebyshev's theorem.** For any set of data, at least $(1-1/k^2)$ of the values lie within $k$ standard deviations of the mean.

**Chi-square.** A probability distribution used here in goodness-of-fit tests and for tests of independence.

**Confidence interval.** The range of values that could be taken at a given significance level.

**Continuous random variable.** One which can assume values associated with a whole line interval.

**Correlation coefficient.** A measure of the relationship between two variables.

**Critical value.** A value used as a threshold to decide whether or not to reject the null hypothesis.

**Descriptive statistics.** Collecting, organizing, and presenting data.

**Discrete random variable.** One which assumes only a countable number of values.

**Empirical rule.** In symmetric "bell-shaped" data, about 68% of the values lie within one standard deviation of the mean, about 95% lie within two standard deviations of the mean, and more than 99% lie within three standard deviations of the mean.

**Expected value.** (or **average** or **mean**) For a discrete random variable $X$, this is the sum of the products obtained by multiplying each value $x$ by the corresponding probability $P(x)$.

**Geometric mean.** The $n$th root of the product of $n$ numbers.

**Harmonic mean.** The reciprocal of the arithmetic mean of the reciprocals.

**Histogram.** A visual representation of data in which relative frequencies are represented by relative areas.

**Inferential statistics.** Analyzing sample data to draw reasonable inferences about population characteristics.

**Interquartile range.** The range of the remaining values after removing the upper and lower one-quarter.

**Mean.** Result of summing the values and dividing by the number of values.

**Median.** The middle number when a set of numbers is arranged in numerical order. If there are an even number of values, the median is the result of adding the two middle values and dividing by two.

**Mode.** The most frequent value.

**Normal distribution.** A particular bell-shaped, symmetric curve with an infinite base.

**Null hypothesis.** A claim to be tested, often stated in terms of a specific value for a population parameter.

**Operating characteristic curve.** A graphical display of $\beta$ values.

**Outlier.** An extreme value falling far from most other values.

**Percentile ranking.** Per cent of all scores which fall below the value under consideration.

**Poisson distribution.** A probability distribution which can be viewed as the limiting case of the binomial when $n$ is large and $p$ is small.

**Population.** Complete set of items of interest.

**Power curve.** The graph of probabilities that a Type II error is not committed.

**Probability.** A mathematical statement about the likelihood of an event occurring.

**Probability distribution.** A listing or formula giving the probability for each value of some random variable.

**P-value.** The smallest value of $\alpha$ for which the null hypothesis would be rejected.

**Random sample.** When the sample is selected under conditions such that each element of the population has an equal chance to be included.

**Range.** The difference between the largest and smallest values of a set.

**Random variable.** Real numbers associated with the potential outcomes of some experiment.

**Regression line.** The best fitting straight line, that is, the line which minimizes the sum of the squares of the differences between the observed values and the values predicted by the line.

**Relative variability.** Quotient obtained by dividing the standard deviation by the mean. Usually expressed as a percentage.

**Sample.** Part of a population used to represent the population.

**Scatter diagram.** A visual display of the relationship between two variables.

**Significance level.** The choice of α-risk in a hypothesis test.

**Simple ranking.** After arranging the elements in some order, noting where in the order a particular value falls.

**Skewed.** A distribution spread thinly on one end.

**Standard deviation.** The square root of the variance.

**Stem and leaf display.** A pictorial display giving the shape of the histogram, as well as indicating the values of the original data.

**Student *t* distribution.** A bell-shaped, symmetrical curve which is lower at the mean, higher at the tails, and more spread out than the normal distribution; often used when working with small samples.

**Trimmed mean.** Taking away an equal percentage of the lowest and highest values before calculating the arithmetic mean.

**Type I error.** The error of mistakenly rejecting a true null hypothesis.

**Type II error.** The error of mistakenly failing to reject a false null hypothesis.

**Variance.** The average of the squared deviations from the mean.

**Z-score.** The number of standard deviations a value is away from the mean.

# INDEX

# BARRON'S COLLEGE REVIEW SERIES

## *Focus on Facts*
## *Summarize Vital Topics*
## *Excel in Your Course*

**BARRON'S COLLEGE REVIEW SERIES** *Science*

### Biology
John Snyder, Ph.D. and C. Leldand Rodgers, Ph.D.

*Papilio machaon*

TOPIC SUMMARIES • KEY TERMS WITH DEFINITIONS
REVIEW QUESTIONS

A survey of biology that covers characteristics and relationships of life, physical and chemical foundations of biology; cells, tissues, organs, and systems; the molecular basis for heredity; biodiversity; much more.

**BARRON'S COLLEGE REVIEW SERIES** *Literature*

### English Literature:
**1800 to 1900**
Arthur H. Bell, Ph.D. and Bernard D.N. Grebanier, Ph.D.

*Dickens*

CHRONOLOGY • WORKS AT A GLANCE • TOPIC SUMMARIES • GLOSSARY • SUGGESTED READINGS • REVIEW QUESTIONS

A survey of nineteenth-century English literature, from the Romantics such as Byron and Keats through the late Victorian novelists and essayists, including Swinburne, Browning, Dickens, George Eliot, Ruskin, Carlyle, and many others.

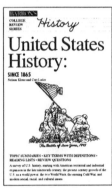

**BARRON'S COLLEGE REVIEW SERIES** *History*

### United States History:
**SINCE 1865**
Nelson Klose and Curt Lader

*The Battle of Iwo Jima, 1945*

TOPIC SUMMARIES • KEY TERMS WITH DEFINITIONS
READING LISTS • REVIEW QUESTIONS

A survey of U.S. history, starting with American territorial and industrial expansion in the late nineteenth century, the present century growth of the U.S. as a world power, the two World Wars, the ensuing Cold War, and modern social, racial, and cultural issues.

**BARRON'S COLLEGE REVIEW SERIES** *Literature*

### World Literature:
**1800 TO THE PRESENT**
Arthur H. Bell, Vincent P. Hopper, and Bernard D.N. Grebanier

*Lincoln*

CHRONOLOGY • WORKS AT A GLANCE • TOPIC SUMMARIES • GLOSSARY • SUGGESTED READINGS • REVIEW QUESTIONS

A survey of major world writing from the dawn of the Romantic movement in the nineteenth century to the present day. It begins with figures such as Rousseau, Goethe, and Schiller, and concludes with major contemporary works including writing from Third World cultures.

---

Each title in *Barron's College Review Series* offers you an overview of a college-level course, and makes a fine supplement to your main textbook.

You'll find topic summaries, lists of key terms, bibliographies, review questions, and more.
**Each book: Paperback, $11.95, Can$15.95**
**(Physics and Statistics are $13.95, Can$17.95)**

---

American Literature:
**1930 to the Present**
*A. Bell, D. W. Heiney, and L. H. Downs*
**ISBN 1836-2**

Biology
*J. Snyder and C. L. Rodgers*
**ISBN 1862-1**

English Literature
**The Beginnings to 1800, 2nd Ed.**
*A. Bell and B. Grebanier*
**ISBN 1775-7**

English Literature
**1800 to 1900, 2nd Ed.**
*A. Bell and B. Grebanier*
**ISBN 1678-5**

English Literature
**1900 to the Present, 2nd Ed.**
*A. Bell, D. W. Heiney, and L. H. Downs*
**ISBN 1837-0**

Physics
*Jonathan S. Wolf*
**ISBN 9522-7**

Statistics
*Martin Sternstein, Ph.D.*
**ISBN 9311-9**

United States History
**To 1877**
*N. Klose and R. F. Jones*
**ISBN 1834-6**

United States History
**Since 1863**
*N. Klose and C. Lader*
**ISBN 1835-4**

World Literature
**Early Origins to 1800, 2nd Ed.**
*A. Bell, V. P. Hopper, and B. Grebanier*
**ISBN 1811-7**

World Literature
**1800 to the Present, 2nd Ed.**
*A. Bell, V. P. Hopper, and B. Grebanier*
**ISBN 1812-5**

---

Books may be purchased at your bookstore, or by mail from Barron's. Enclose check or money order for the total amount plus sales tax where applicable and 10% for postage and handling charge (minimum charge $3.75, Canada $4.00). Prices subject to change without notice.

**Barron's Educational Series, Inc.**
250 Wireless Blvd. • Hauppauge, NY 11788
**In Canada:** Georgetown Book Warehouse
34 Armstrong Ave., Georgetown, Ont. L7G 4R9

ISBN Prefix: 0-8120
$ = U.S. Dollars
Can$ = Canadian Dollars

R 7/96 (#58)